Frontiers in Electrical Engineering
Vol. 1

Active-Matrix Organic Light-Emitting Display Technologies

Authored By

Shuming Chen & Jianning Yu

Department of Electrical and Electronic Engineering, South University of Science and Technology of China, Shenzhen, Guangdong 518055, P. R. China

Yibin Jiang, Rongsheng Chen & Tsz Kin Ho

Department of Electronic and Computer Engineering, the Hong Kong University of Science and Technology, Hong Kong, P. R. China

advertisements or ideas contained in the Work.

Limitation of Liability:

In no event will Bentham Science Publishers, its staff, editors and/or authors, be liable for any damages, including, without limitation, special, incidental and/or consequential damages and/or damages for lost data and/or profits arising out of (whether directly or indirectly) the use or inability to use the Work. The entire liability of Bentham Science Publishers shall be limited to the amount actually paid by you for the Work.

General:

1. Any dispute or claim arising out of or in connection with this License Agreement or the Work (including non-contractual disputes or claims) will be governed by and construed in accordance with the laws of the U.A.E. as applied in the Emirate of Dubai. Each party agrees that the courts of the Emirate of Dubai shall have exclusive jurisdiction to settle any dispute or claim arising out of or in connection with this License Agreement or the Work (including non-contractual disputes or claims).
2. Your rights under this License Agreement will automatically terminate without notice and without the need for a court order if at any point you breach any terms of this License Agreement. In no event will any delay or failure by Bentham Science Publishers in enforcing your compliance with this License Agreement constitute a waiver of any of its rights.
3. You acknowledge that you have read this License Agreement, and agree to be bound by its terms and conditions. To the extent that any other terms and conditions presented on any website of Bentham Science Publishers conflict with, or are inconsistent with, the terms and conditions set out in this License Agreement, you acknowledge that the terms and conditions set out in this License Agreement shall prevail.

Bentham Science Publishers Ltd.
Executive Suite Y - 2
PO Box 7917, Saif Zone
Sharjah, U.A.E.
Email: subscriptions@benthamscience.org

**BENTHAM
SCIENCE**

CONTENTS

The Cover Image has been provided by Shuming Chen, the author of this book

PREFACE

Over the last several decade years, great successes have been achieved in the industrialization of organic light-emitting diode (OLED) technologies. Active-matrix OLED (AMOLED) emerged as an important and low-cost candidate to replace liquid crystal displays due to its attractive advantages like self-emitting, high efficiency, high contrast, vivid color, fast response, and flexible form factor. The purpose of this eBook is to present an introduction to the subject of AMOLED and their related technologies which are generally integrated to the production application, including OLED basic working principles, fabrication and characterization, white OLED technologies, light outcoupling technologies, encapsulation technologies, thin film transistor backplane technologies, driving scheme, circuit and layout design technologies. Although it is impossible to cover completely the vast amount of publications concerning these topics, we will select those key areas in device structures, fabrication techniques and application that we feel are most pertinent to the practical production.

This book will be helpful for young scientists and engineers who work at the development of practical OLED display and OLED lighting. Target readers include researchers in organic electronics field, undergraduate and graduate students who study in OLED display and OLED lighting, and also for the engineers who work in OLED industry. Through reading the eBook, readers can get a comprehensive and insight view of the AMOLED display, including the principles of OLED, the fabrication and characterization techniques of OLED and various related technologies for display system integration, which is benefit for their future academic or industrial career.

Shuming Chen
Department of Electrical and Electronic Engineering
South University of Science and Technology of China
Shenzhen, Guangdong 518055, P. R. China

LIST OF CONTRIBUTORS

Shuming Chen   Department of Electrical and Electronic Engineering, South University of Science and Technology of China, Shenzhen, 518055, P. R, China

Jianning Yu   Department of Electrical and Electronic Engineering, South University of Science and Technology of China, Shenzhen, 518055, P. R, China

Yibin Jiang   Department of Electronic and Computer Engineering, Hong Kong University of Science and Technology, Hong Kong

Rongsheng Chen   Department of Electronic and Computer Engineering, Hong Kong University of Science and Technology, Hong Kong

Tsz Kin Ho   Department of Electronic and Computer Engineering, Hong Kong University of Science and Technology, Hong Kong

CHAPTER 1

Introduction to Organic Light-Emitting Display Technologies

Shuming Chen[*], Jianning Yu

Department of Electrical and Electronic Engineering, South University of Science and Technology of China, Shenzhen, 518055, P. R. China

Abstract: Organic Light Emitting Diodes (OLEDs) emerged 30 years ago as a very promising flat-panel display technology because of their lots of advantages. OLEDs consist of several nanometer-scale thin films between anode and cathode, which can be integrated on various substrates, especially the ultra-thin and flexible substrates. As a result, OLEDs present advantages of high efficiency, lightweight, fast response time, wide viewing angle, vivid color, high contrast and *etc*. After development of technology improvement over past several decades, enormous progress has been made to promote OLED products to be launched into market. In this chapter, we review the development history of OLEDs, with special emphasis on several key technical innovations which significantly advanced the industrialization. From two main application filed of display and lighting, we describe the working mechanism, technical development, advantages and disadvantages, and production status of OLED technology.

Keywords: Composite cathode layer, Display, Electrical doping, Electroluminescence (EL), Emitter doping, Flexible OLED, ITO surface treatment, Light outcoupling, Lighting, Multilayer structure, Organic light-emitting diode (OLED), Phosphorescent emitter, Thermally activated delayed fluorescence.

INTRODUCTION

Avatar wakes up on a sunny weekend morning. As he washes, the mirror displays today's weather and his agenda. He makes a coffee for himself and sits down at

[*] **Corresponding Author Shuming Chen:** Department of Electrical and Electronic Engineering, South University of Science and Technology of China, Shenzhen, 518055, P. R. China; Tel.: +86-755-88018522; E-mail: chen.sm@sustc.edu.cn

the table; today's latest news is already displayed on the table. He rolls up his laptop and puts it into his bag. As he walks down the street, the sun starts charging his bag and his laptop, PDA *etc* inside the bag. Night is coming; the windows shine the light and light up the whole room.

These dream displays/lightings usually appear in fictions or movies. However, they will become a reality in the near future with the help of organic optoelectronics devices like organic light-emitting diodes (OLEDs) [1 - 4], organic photovoltaic cells (OPVs) [5 - 7] and organic thin film transistors (OTFTs) [8, 9]. Being organic-based devices, they can be flexible as cloth or transparent as glass, thus opening a new wide variety of applications. This chapter provides a basic background to OLED technology. After a brief introduction of the OLED history, an overview of the working mechanisms of OLEDs is presented. The fabrication and characterization of OLEDs is then given, followed by outlining their applications.

DEVELOPEMNT HISTORY OF OLEDS

Organic electroluminescence (EL) is emission of light from organic materials in response to electric current, which was found for the first time in a thick (50 μm to 1 mm) anthracene crystal by Pope's group from New York university in 1960s [10]. However, they did not receive too much interest due to their high driving voltage (>100 V) and weak EL emission in those early crystal devices [10, 11]. In 1987, Tang and VanSlyke, who were both from Eastman Kodak, introduced the first double layer thin-film OLED, which consisted of 60 nm Alq_3 and 75 nm diamine sandwiched between a Mg:Ag reflective cathode and an indium-tin-oxide (ITO) transparent anode as shown in Fig. (**1.1**); all are deposited by vacuum thermal sublimation. The resultant device, driven at 10 V, exhibited a luminance of 1000 cd/m^2 with a moderate external quantum efficiency of 1% (luminous efficiency 1.5 lm/W) [1]. Shortly afterwards, in 1990, a research group from Cambridge University announced the first polymer-based LED prepared by spin-coating the conducting conjugated polymer poly(p-phehylene vinylene) (PPV) onto the ITO-coated glass substrate, followed by capping a thermal evaporated Al cathode. The external quantum efficiency of the device is 0.5% mainly due to the poor exciton confinement in such a single layer device [12].

From then on, OLED has attracted considerable research attention due to its potential applications in flat-panel-display and solid-state-lighting. Enormous progress has been made in the past 30 years in the improvement of driving voltage, efficiency, luminance, color saturation and stability. Several key technologies that significantly advance the development of OLEDs are listed below.

1. Emitter doping. In 1989, Tang, VanSlyke and Chen introduced the concept of emitter doping [2]. By doping highly fluorescent molecules into the Alq_3 matrix, a 2~3 folds efficiency improvement has been achieved. Also, it was observed that the reliability of the doped devices was significantly improved due to the elimination of intermolecular hydrogen bonding between the dopant molecules [13, 14]. More importantly, by simply changing the guest dopants, the emission color can be readily tuned. The doping technique is widely used in modern high efficiency devices [21, 27, 28].

2. Multi-layer structure. In 1990, Adachi *et al.* introduced a double-heterojunction structure with a 5 nm emission layer enclosed by an electron-transport layer and a hole-transport layer [15]. With such configuration, the charge carriers and molecular excitons are better confined within the emission layer, thus preventing charge carrier leakage and/or molecular exciton quenching by the electrodes. Inspired by this pioneering work, modern devices typically adopt a multi-layer structure with each layer specially functioning as hole-injection, hole-transport, electron-block, light emission, hole-block, electron-transport or electron-injection.

3. Al/LiF composite cathode layer. In 1996, Huang *et al.* discovered that the presence of an ultra thin (0.5 nm) LiF layer between the organic layer and Al layer significantly enhances electron injection [16]. Also, the Al/LiF has better chemical stability against atmospheric corrosion compared to the MgAg alloy electrode. The improved electron injection probably results from the release of low work function Li atoms and subsequent formation of the Alq_3 radical anions. Today, LiF is a necessity for making efficient bottom-emitting OLEDs.

4. ITO surface treatment. In 1997, Wu *et al.* demonstrated that the chemical composition of ITO surfaces could be substantially modified by O_2 plasma treatment [17]. The work function of the treated ITO surfaces was increased by 100-300 meV compared to the cleaned as-grown ITO surfaces, mainly due to

the increase of surface oxygen concentration. As a consequence, hole injection ability of the treated ITO is enhanced dramatically, leading to a reduction of driving voltage and an improvement of efficiency, luminance and reliability. Later on, CF_4 plasma treatment was also found to be an effective way to enhance the hole injection ability of ITO [18]. The ITO surface treatment is one of the standard procedures for making bottom-emitting OLEDs today.

5. Phosphorescent emitter. Light emission in fluorescent small-molecular OLEDs occurs as a result of the radiative decay of singlet excitons, which is only 25% of the total excitons according to spin statistics. For phosphorescent materials, by introducing heavy metal atoms into organic molecules to increase the spin-orbit coupling, the singlet and triplet excitonic states can be mixed, thereby making the radiative recombination of all excitons allowed. The phosphorescent emission was first observed in 1998 by Baldo *et al.* in Forrest's group from Princeton University [19, 20]. A nearly 100% internal quantum efficiency in phosphorescent OLEDs was subse-quently demonstrated [21].

6. Electrical doping. Because the carrier mobility of most organic materials is low, the driving voltage of the OLEDs typically in the range of 3-10 V is substantially higher than that of their inorganic counterparts. To low down the driving voltage and hence the power consumption, Kido first proposed using metals with low work function such as Sm, Sr and Li doped with the electron-transport materials to increase the conductivity of the electron-transport materials in 1998 [22]. Leo's group from Dresden University further proposed a more elaborate p-i-n structure. In such structure, the hole-transport and electron-transport materials are doped by suitable molecular (F4-TCNQ) or metal (Li) dopants [23]. Due to electron transfer from the highest occupied molecular orbital (HOMO) of the hole-transport materials to the lowest occupied molecular orbital (LUMO) of the dopant materials (p-doped) or from the HOMO of dopant materials to the LUMO of electron-transport materials (n-doped) [24], the charge carrier density of the transport materials is increased dramatically, resulting in an enhanced conductivity. With the p-i-n structure, extremely low driving voltage of 2.55 V for 100 cd/m^2 and 2.9 V for 1000 cd/m^2, approximately cor-responding to the photon energy of the emitter, has been demonstrated [25].

7. Light outcoupling. Classical ray optics predicts that only about 20% of the

internal emission can be escaped from the substrate and the remaining 80% photons are trapped inside the devices because of the total internal reflection (TIR) at the interface of glass/air and ITO/glass [26]. By using light out-coupling techniques, the trapped photon can be effectively extracted and thus the efficiency can be greatly improved by a factor of 1.2-2.3, which have been experimentally proved. For example, a high refractive index substrate can be used to effectively extract the ITO/organic waveguide light [27, 28], while a microlens array on the back side of the substrate can be employed to outcouple the substrate waveguide light [29, 30]. A very high efficiency of 210 lm/W has been demonstrated very recently by using a high refractive index substrate equipped with a mocrolens array, which has been improved by a factor of 2.3 compared with the devices without applying the light outcoupling techniques [28].

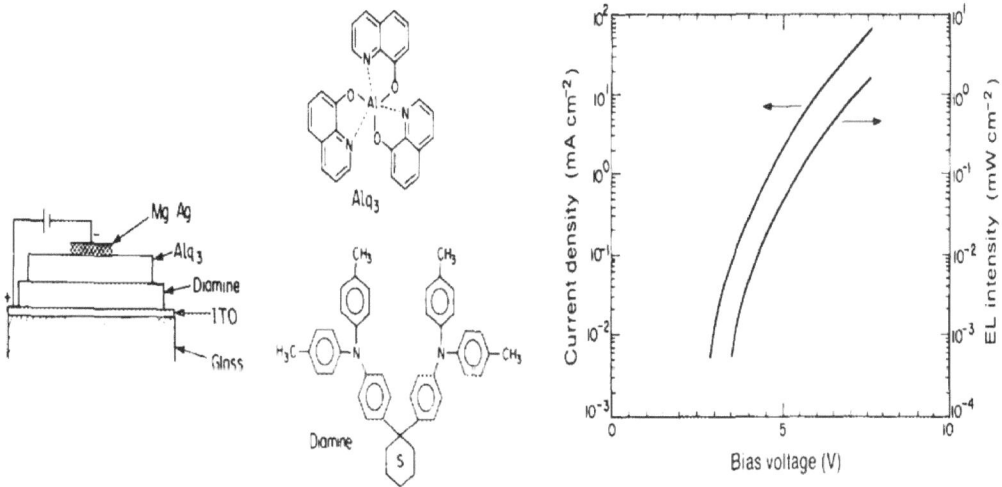

Fig. (1.1). Device structure of the first reported OLED (left), molecular structures of the materials used (middle) and the J-V-L characteristics of the devices (right). [1]

These key fundamental technologies are widely employed for fabricating high performance devices nowadays. Some of them even become the standard and essential procedures for obtaining efficient devices. For example, Leo *et al.* demonstrated a white light OLED exhibiting a high efficiency of 90 lm/W at 1000 cd/m^2 recently by combining the aforementioned techniques 1, 2, 5, 6 and 7 [27].

BASIC PHYSICS OF OLEDS

When a DC voltage is biased to the electrodes of the OLEDs, light emits immediately. Take the single layer devices as shown in Fig. (**1.2**) for example, the light generation process usually includes the below five successive steps [33]. (1) Charge injection to the organic layers from the electrodes, (2) carriers transportation within the organic layers under the driven of an external electric filed, (3) molecular excitons (hole-electron pairs) formation due to the Colombian interaction of holes and electrons, (4) exciton radiaitive decay leading to the generation of photons and (5) photons escape from the structure through the (semi-)transparent electrode. Each of these five steps should be thoroughly optimized so that a high efficiency can be obtained.

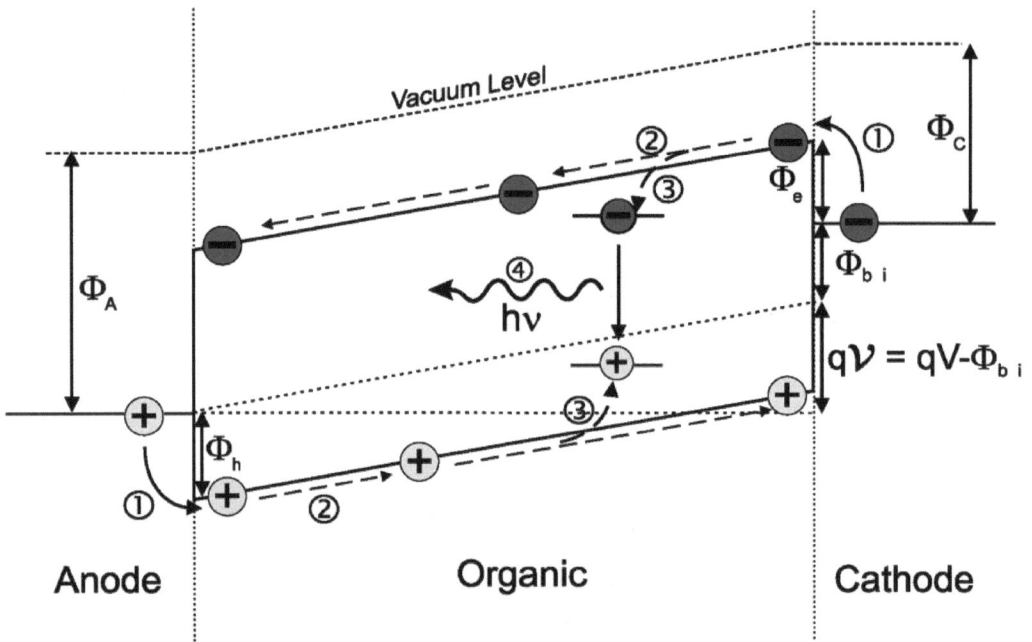

Fig. (1.2). Basic steps of EL in a single layer device [33].

Charge Carriers Injection

The operating voltage and luminance efficiency of OLEDs are largely determined by the charge injection and transport. More organic materials have a relative wide energy band-gap (2-3 eV) and therefore the intrinsic carrier concentration generated by thermal excitation is generally negligible ($<10^{10}$ /cm^3) and from this view point the materials are traditionally regarded more as insulators instead of as semiconductors [33]. Most of the carriers that are used to generate photons therefore have to be extracted from the electrodes. Injection of carriers from the electrodes (anode and cathode) requires that carriers overcome or tunnel an energy barrier at the organic-electrode interface. The height of the barrier is generally dependent on the difference of the LUMO/HOMO of the organic carrier-transport materials and the work function of the electrode. Two injection models, namely the Richardson Schottky (RS) thermionic emission model and the Fowler-Nordheim (FN) tunneling model are often employed to explain the charge carrier injection process at low electric filed [32]. Fig. (**1.3**) schematically shows the two injection mechanisms of electrons moving into the organic layers from the cathode.

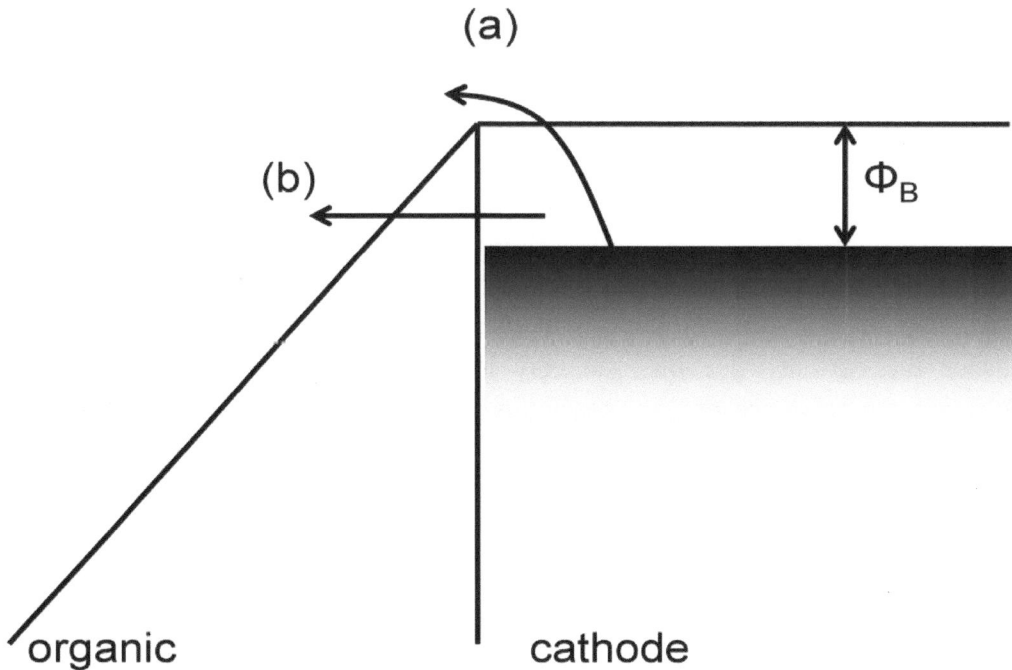

Fig. (1.3). Injection mechanisms of electrons into organic layers from the cathode driven by voltage V: (a) direct tunneling; and (b) thermionic emission.

In the FN tunneling model, the electrons are injected from the Fermi level of the cathode into the organic layers by tunneling through a thin triangle barrier. The injected current density is given by:

$$J_{FN} = \frac{A^*}{\Phi_B}\left(\frac{qF}{\alpha k_B}\right)^2 \exp\left(-\frac{2\alpha\Phi_B^{3/2}}{3qF}\right) \propto F^2 \exp\left(-\frac{2\alpha\Phi_B^{3/2}}{3qF}\right)$$ .... (1.1)

with the Richardson constant $A^* = \frac{4\pi qm^* k_B^2}{h^3}$, external field $F = \frac{V}{d}$, $\alpha = \frac{4\pi\sqrt{2m^*}}{h}$, ($k_B$ is the Boltzmann's constant; q is the elementary charge; m^* is the effective electron mass; h is the Planck's constant; d is the organic layer thickness). Plotting of $\ln(J/F^2)$ *versus* $1/F$ would produce a straight line, and the slope of the curves allows one to determine the injection barrier Φ_B.

The RS thermionic emission take account the influence of image charge potential which is lowered by electric field F. Based on this consideration, the injected current density is given by:

$$J_{RS} = A^* T^2 \exp\left(-\frac{\Phi_B}{k_B T}\right) \exp\left(-\frac{\beta_{RS} F^{1/2}}{k_B T}\right), with \ \beta_{RS} = \sqrt{\frac{q^3}{4\pi\varepsilon\varepsilon_0}}$$ .... (1.2)

From equations (1.1) and (1.2), one can see that both mechanisms indicate that the current injection is strongly dependent on the potential barrier Φ_B. Hence, to enhance the current injection, one may lower the Φ_B. In the simplest case neglecting the existence of the interface dipole induced by a variety of origins such as interface state, surface rearrangement, chemical bonding, charge transfer and *etc.*, the barrier height is dependent on the energy level difference between the HOMO/LUMO of the organic semiconductors and the work function of the electrodes. Hence, for realizing efficient injection, electrode materials with appropriate work function to match the HOMO/LUMO of the organic materials should be chosen. Al and Ag are widely employed as the electrode materials for the OLEDs, due to their chemical stability and high reflectivity in the visible light region [16, 33]. However, both of them have a relatively high work function of ~4.3 eV, while the LUMO value of typically electron-transport materials (Alq_3,

TPBi, Bphen *etc.*) range from 2 to 3 eV. Apparently, a high electron injection barrier exists when Al or Ag is served as the cathode. Considerable efforts have been devoted in the past several decades to enhance the injection of electron from Al or Ag to the organic materials. In Tang's revolutionary work, an alloy electrode with 90% relatively low work function Mg doped with 10% stable Ag is employed for simultaneously improving the chemical stability and the electron injection capability of the cathode [1]. Later on, ultrathin alkaline fluorides such as LiF [16], CsF [34] and NaF [35] inserted between the Al layer and the electron-transport layer were found to facilitate the electron injection greatly. A composite electrode structure with a low work function reactive metal like Ca [36], Ba [37] and Yb [38] protected by a stable metal such Al, Ag or Au was also reported. Doping the electron-transport materials with Li [22], Cs [39], Li_2CO_3 [40], Cs_2CO_3 [41] was found to dramatically increase the conductibility of the electron-transport materials, resulting in a significantly enhanced electron injection current.

Transparent ITO with a work function of 4.7 eV is commonly employed as the anode for injecting holes to the HOMO of the organic hole-transport materials. Various surface treatment methods such as UV ozone [42], O_2 [17] or CF_4 [18] plasma are employed to increase the work function of ITO. A remarkably enhanced hole injection from plasma treated-ITO to NPB was observed due to the matching of surface work function of ITO with the HOMO (~5-5.5 eV) of NPB [17, 18]. Alternatively, thin buffer layer like WO_3 [43, 44], MoO_3 [45 - 48], V_2O_5 [49, 50] inserted between the ITO electrode and NPB was also found to dramatically enhance the hole injection. The working mechanisms of these buffer layers are still under debate. Many hypotheses like surface band bending, interface dipole formation and charge transferring have been proposed to explain the working mechanism. The hole injection can be enhanced more significantly by doping NPB with WO_3, MoO_3 or V_2O_5 [51 - 55]. Because the charges are transferred from the HOMO of NPB to the guest acceptors, the hole conductibility of the blending film is considerably higher than that of the neat NPB film. No matter what kind of methods were used, the ultimate purposes are the same, to realize an Ohmic contact. It was reported by doping the hole and electron-transport layers thus forming a p-i-n structure, an Ohmic contact can be achieved, resulting in extremely low driving voltage of 2.55 V for 100 cd/m^2 in a green OLED [25].

Charge Carriers Transportation

After injection from the electrodes, the charge carriers are transported through materials driven by an applied electric filed. In contrast to inorganic semi-conductors, where carriers move in well-defined bands, the charge movement in organic semiconductors is a stochastic process of hoping between localized states, resulting in a very low mobility typically ranging from10^{-5} to 10^{-3} cm^2/Vs for the holes and 10^{-6} to 10^{-4} cm^2/Vs for the electrons. In many cases the mobility is Poole-Frenkel field dependent given by [56]:

$$\mu(F) = \mu_0 \exp\left(\sqrt{F/F_0}\right)$$ .... (1.3)

where μ_0 is the zero filed mobility and F_0 is the characteristic filed [56]. When more charges are injected to a semiconductor compared to the number of charges that materials have in the thermal equilibrium state, space charges will be formed by the excess injected charges, which induces an internal electric field with its direction opposite to the external electric field and thus weaken the injection of the charge from the electrode in turn. In this case, the current is limited by the bulk of the semiconductor instead of by the injection from the electrode, that is, by the rate at which the carriers are drifted out of the space charge region. Neglecting the traps, the space charge limited current obeys the Mott-Gurney law [57]:

$$J_{SCL} = \frac{9}{8}\varepsilon\mu\frac{V^2}{d^3}$$ .... (1.4)

Where d represents the thickness of the organic layers. The mobility of the charge carrier can be simply determined by using this equation to fit the experimental J-V curve. The SCLC law shows that the current density increases proportionally to the square of the external driving voltage. However, in real case, due to the presence of traps, at low driving voltage, the current is usually lower than the SCLC value. The traps are originated from the residual impurities and/or from the disorder of molecular conformation [58]. These trapped charges generally cannot move and thus do not contribute to the current. At very small voltage, almost no charges are injected from the electrodes, and thus the current is determined by the movement of the thermally generated free charges in the bulk. In an Ohmic conduction mode, the current density is governed by [59]:

$$J = \frac{e\mu n_0 V}{d}$$ .... (1.5)

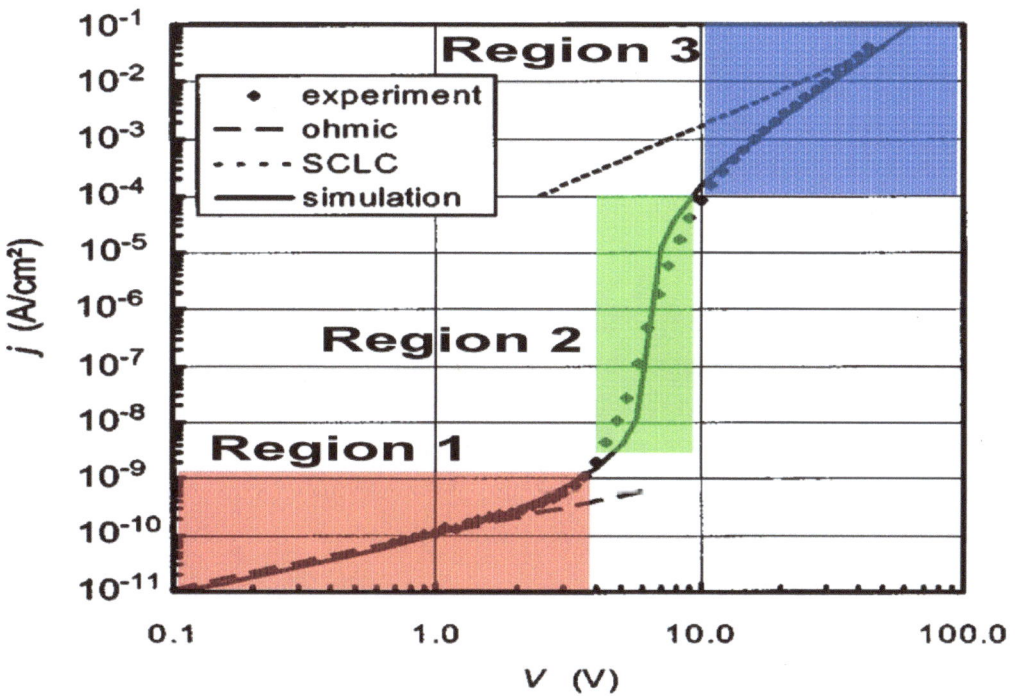

Fig. (1.4). Current density in a single layer hole-only device with structure ITO/m-MTDATA (850 nm)/Ag. Region 1: Ohmic conduction current; region 2: trap-limited current; region 3: trap-free space charge limited current [60].

where n_0 is the density of the charges generated by thermally excitation. Thus, the current is lineally proportional to the driving voltage, as reflected in region 1 in Fig. (**1.4**). As driving voltage increases, more traps are gradually filled; as a result, the density of empty trap states is reduced, which consequently increases the effective carrier mobility $\mu_{eff} = \mu\left(\dfrac{n_{inj}}{n_t}\right)$. Here n_t is the density of total trapped charge and n_{inj} is the density of injected charge. Therefore, the current increases dramatically and the current-voltage characteristic obeys a higher power law, as shown in region 2 in Fig. (**1.4**). If considering the traps are exponentially distributed in the gap, then the trap charge limited (TCL) current is given by [31, 59]:

$$J_{TCL} = N_C \mu q \left(\frac{\varepsilon\varepsilon_0 l}{N_t q (l+1)}\right)^l \left(\frac{2l+1}{l+1}\right)^{l+1} \frac{V^{l+1}}{d^{2l+1}}$$ .... (1.6)

where $l = \dfrac{E_t}{k_B T}$ characters the trap distribution and N_C is the density of states at the HOMO or LUMO level. A $l \approx 6$ to 8 is often observed. At sufficient high injection level, the transport of electrons is no longer influenced by the traps as they are completely filled. An ideal SCLC behavior as shown in region 3 in Fig. (**1.4**) is expected.

In real devices where recombination and hole and electron carrier injection occurs, the situation becomes more complicated. A complete device model is required to include charge injection, transportation and recombination as well as the effect space charge. In this model, transport of electron and hole is described using a drift-diffusion or drift current and the continuity equations, coupled to Poission's equation [61]:

$$\frac{dn(x)}{dt} = \frac{1}{q}\frac{dj_n(x)}{dx} - R(x)p(x)n(x)$$ .... (1.7)

$$j_n(x) = q\mu_n(x,E)n(x)E(x) + qD(\mu_n)\frac{dn(x)}{dx}$$ .... (1.8)

$$\frac{dE}{dx} = \frac{q}{\varepsilon\varepsilon_0}\left[p(x) - n(x)\right]$$ .... (1.9)

where p, n is the density of holes and electrons, respectively; E, D is the electric field and the diffusion constant, respectively; R is the recombination constant given by:

$$R = \frac{q}{\varepsilon \varepsilon_0}\left(\mu_n + \mu_h\right)$$

.... (1.10)

Exciton Formation and Recombination

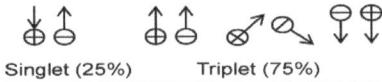

When the distance between a hole and an electron is short sufficiently so that their mutual Coulomb binding energy is larger than the thermal energy $K_B T$, a binding hole-electron pair so called exciton is created with a recombination rate given by equation (1.10). The excitons undergo radiative decay if their diffuse without quenching, and consequently generates an internal emission S:

$$S \propto \int_0^L \frac{s(x)}{\tau_e} dx$$

.... (1.11)

where s(x) is the density of the excitons; L is the exciton diffusing length and τ_e is its lifetime.

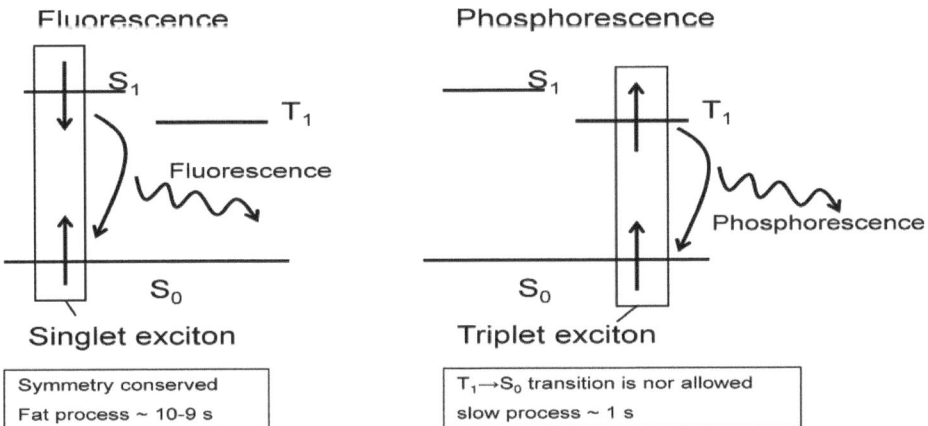

Fig. (1.5). A: Possible spin combinations of excitons; and **B**: fluorescence and phosphorescence [63].

Based on the spin combination of a hole and an electron, there are two kinds of exciton, namely singlet exciton and triplet exciton. By definition, the spin directions of the electron-hole of the singlet excitons are antisymmetric, while those of triplet excitons are symmetric [62]. As there is only one antisymmetric spin combination among the four different possible spin combinations as shown in Fig. (**1.5**), the singlet-to-triplet ratio is hence 1:3.

Based on Pauli Exclusion Principle, excited singlet states is allowed to decay, while the decay of triplet states to the ground state is forbidden unless the spin direction is changed during transition. Consequently, the 25% singlet excitons decay fast ($k \sim 10^9$/s) and efficiently due to spin conservation, resulting in an EL emission called fluorescence, while the remaining 75% triplet exctions decay so slowly ($k < 10^6$/s) that most of the energy is lost to non-radiative processes, leading to an extremely weak EL emission called phosphorescence [62, 63].

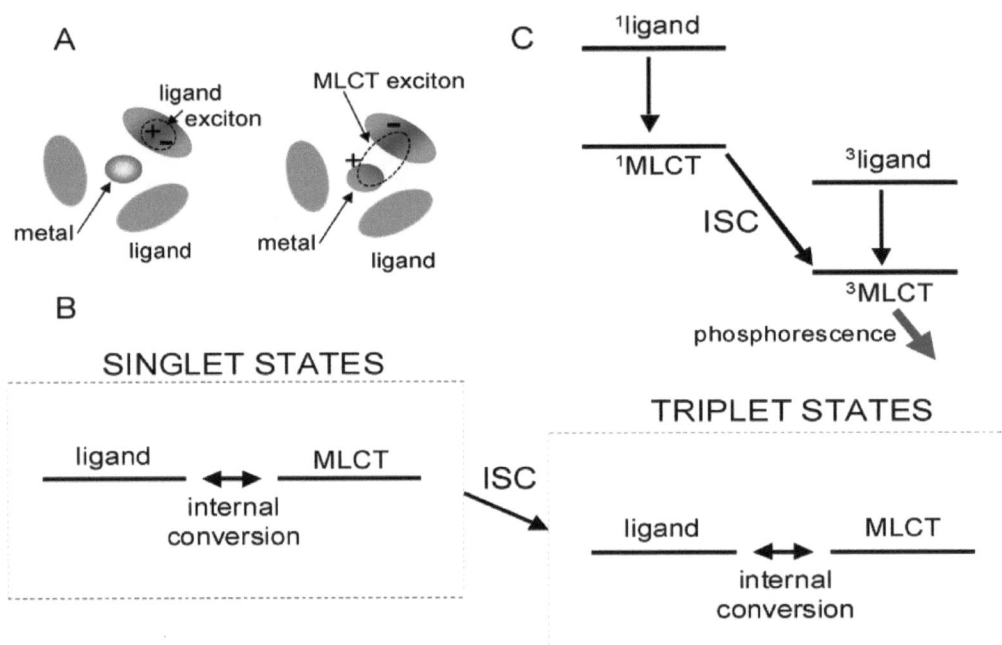

Fig. (1.6). **A**: ligand exciton and MLCT exciton; **B**: ligand to MLCT energy transfer *via* internal conversion and singlet to triplet transfer *via* intersystem crossing are effective for strong spin-orbit coupling; **C**: MLCT triplet excitons decay to the ground state *via* ISC resulting in strong phosphorescence emission [62].

To harvest the triplet excitons, a heavy metal like Pt, Ir or Os is introduced to the organic molecules to increase the spin-orbit coupling, so that the mixing of singlet and triplet excitonic states is achieved [62]. In such mixing states, the triplet gains some singlet character and thus can decay to the ground state efficiently. In the metal-organic complexes, the emissive state is generally composed of a metal-ligand charge transfer (MLCT) exciton and a ligand-centered exciton as shown in Fig. (**1.6A**). As the MLCT state overlaps with the heavy-metal atom, it has superior singlet-triplet mixing. For strong spin-orbit coupling, singlet to triplet energy transfer *via* intersystem crossing (ISC) and ligand to MLCT energy transfer *via* internal conversion are extremely fast (Fig. **1.6B**). For high performance phosphorescence system as shown in Fig. (**1.6C**), the MLCT triplet state generally has the lowest energy, thus minimizing the triplet lifetime and triplet-triplet annihilation due to the strong singlet-triplet mixing of the MLCT state. The subsequent decay of MLCT triplet excitons to the ground state *via* ISC result in strong phosphorescence emission.

To achieve an efficient exciton radiative decay, the emitters are often doped into the charge transport host materials. The emitter doping is one of the key technologies that advance the development of OLED. There are several reasons for adopting the host-guest system. First, the host-guest system provides a facile way to tune the emission color by simply doping the host with suitable color dopants [2, 13]. In addition, the hosts usually possess higher charge mobility than the dopants, thus offering a channel for effective and fast exciton recombination. More importantly, luminescent molecules generally suffer from concentration quenching due to intermolecular π-π interaction, a notorious effect termed aggregation-caused quenching. It is thus necessary to distribute the luminescent molecules to the host matrix to reduce the undesirable intermolecular π-π interaction, so that a high efficiency can be obtained.

However, there are some disadvantages of the host-guest system in terms of performance repeatability and fabrication complexity. The efficiency of the doped devices is doping concentration dependent, however it is difficult to maintain the same doping concentration for different batch devices, thus the efficiency of the devices varies from batch to batch. Also, the fabrication process of the doped film is relatively complex than that of the neat film because the deposition rate of the host and guest should be fine-tuned and steadily maintained during co-deposition to achieve uniform doping profile.

In the host-guest system, the excited host molecules (donors) transfer their energy to the un-excited luminescent dopants (acceptors), which consequently are promoted to an excited state. The subsequent radiative decay of the excited acceptors leads to the generation of the EL emission. Such energy transfer process can be expressed as [63]:

where hv_T and hv_A are the thermal and photon energies emitted in the process. Based on the interaction of $D*$ and A, there are two energy transfer mechanisms namely Förster and Dexter transfer. As shown in Fig. (**1.7A**), in the Förster transfer process, the energy of the excited donor is nonradiatively transferred to the acceptor *via* dipole-dipole coupling.

$$D^* + A \rightarrow D + A^* + hv_T \rightarrow D + A + hv_A \dots\dots\dots\dots\dots\dots (1.12)$$

The range of transfer may extend to 10 nm, if both transitions on the donor and the acceptor are allowed; thus typical doping concentration for effective Förster transfer is ~1% [2, 13], which approximately corresponds to a dopant distance of 100 Å. The Förster transfer plays a dominant role for the transfer of singlet excitons. However, the Förster energy transfer is not possible if one of the transitions is disallowed; for instance, it is impossible to excite the triplet state of the acceptor *via* the Förster mechanism.

The transfer of triplet state to the acceptor can be achieved by electron exchange between the donor and the acceptor, a process called Dexter energy transfer. In contrast with the long range dipole-dipole interaction in the Förster transfer, the electron exchange in the Dexter transfer is a short-range process, which occurs only over a short distance of 10 Å.

Thus the optimal doping concentration for effective Dexter transfer is relatively higher than that for Förster transfer, with a typical value ranging from 6% to 12% [19 - 21]. The energy transfer rate of both mechanisms is strongly dependent on the spectral overlap between absorption spectrum of guest material and PL spectrum of host material. Large spectral overlap results in a faster transfer rate.

A Dipole-dipole coupled (Förster) energy transfer

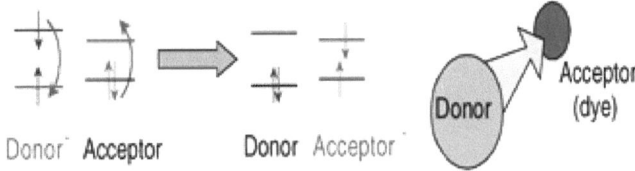

Donor   Acceptor   Donor   Acceptor

**Exciton nonradiatively transferred by
dipole - dipole coupling if transitions are allowed.**   up to ~100 Å

B Electron exchange energy transfer

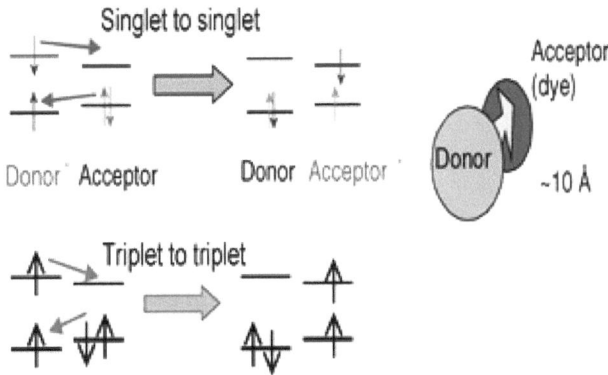

Singlet to singlet

Donor   Acceptor   Donor   Acceptor   ~10 Å

Triplet to triplet

Exciton hops from donor to acceptor.

C

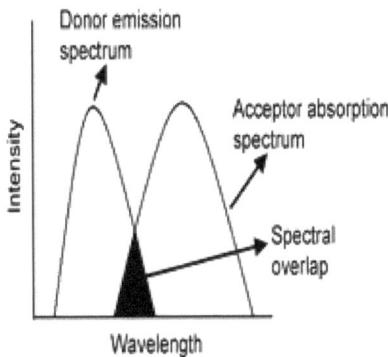

Fig. (1.7). (**A**) Förster and (**B**) Dexter energy transfer. (**C**) The transfer rate is proportional to the spectral overlap between absorption spectrum of acceptor and emission spectrum of donor.

Light Extraction from Devices

Theoretically, all injected electrons can be used to generate photons if perfect charge confinement is achieved and phosphorescent emitters are employed. Nevertheless, only a few of those internally generated photons can escape from the structure while others are trapped inside the devices because of the total internal reflection (TIR) at the interface of ITO/glass and glass/air. The schematic ray diagram for a planar OLED is shown in Fig. (**1.8**). The refractive index for the glass, ITO and organic is around 1.52, 1.9 and 1.7, respectively. Hence, the critical angles at the ITO/glass and glass/air are

$$\theta_1 = \arcsin\left({n_{air}}/{n_{org}} \right) = 36°, \theta_2 = \arcsin\left({n_{sub}}/{n_{org}} \right) = 63° \quad \text{respectively. According to}$$

the ray diagram, the emitted light can be classified into three modes [64]:

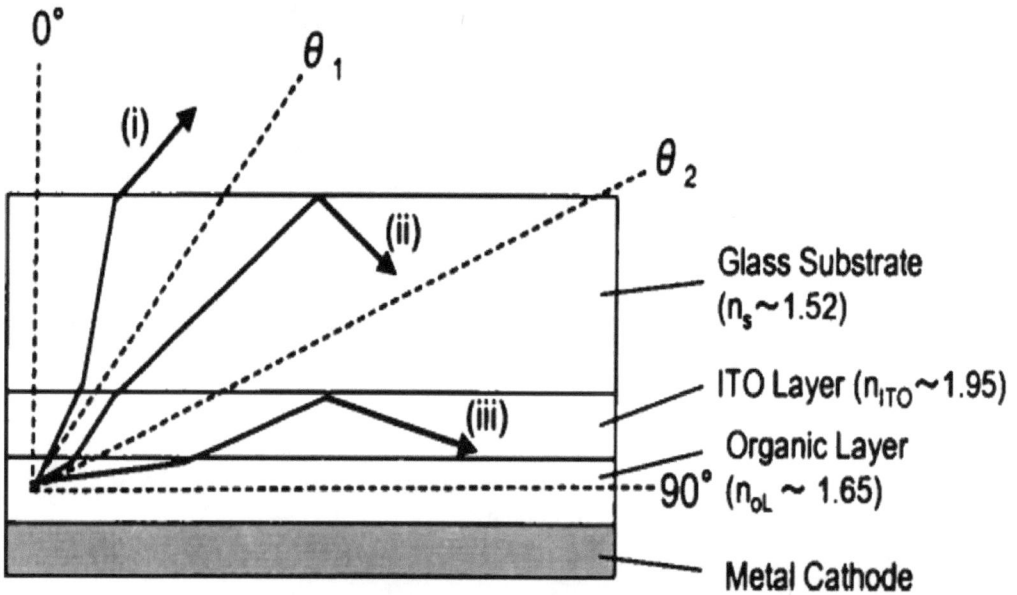

Fig. (1.8). Ray diagram of OLEDs: (**i**) external mode (0∘<theta<theta1), (**ii**) substrate waveguided mode (theta1<theta<theta2) and (**iii**) ITO/organic waveguided mode (theta2<theta<90 °) [64].

1. In the external mode, light with angle $0 \leq \theta < \theta_1$ is escaped from the devices. If the cathode is a prefect reflector and the dipole emission is isotropic, the fraction of generated photons that escapes from the substrate is:

$$\eta_{coupl} = \frac{\int_0^{\theta_1} \sin\theta d\theta}{\int_0^{\pi/2} \sin\theta d\theta} = 1 - \cos\theta_1 = 1 - \sqrt{1 - \frac{1}{n_{org}^2}} \approx \frac{1}{2n^2} = 19.1\%$$

.... (1.13)

2. In the substrate mode, light with angle $\theta_1 \leq \theta < \theta_2$ is trapped and waveguided by the glass substrate and eventually emit from the edges of the substrate. The fraction of generated photons that is waveguided by substrate is:

$$\eta_{sub-wg} = \frac{\int_{\theta_1}^{\theta_2} \sin\theta d\theta}{\int_0^{\pi/2} \sin\theta d\theta} = \cos\theta_1 - \cos\theta_2 = 35.5\%$$

.... (1.14)

3. In the ITO/organic mode, light with angle $\theta_2 \leq \theta \leq 90°$ suffer the TIR that is presented at the ITO/substrate interface, and hence they are trapped inside the ITO/organic and will be quickly absorbed by ITO, organic and metal cathode layers. The fraction of ITO/organic waveguide light is:

$$\eta_{ITO/org-wg} = \frac{\int_{\theta_2}^{90°} \sin\theta d\theta}{\int_0^{\pi/2} \sin\theta d\theta} = \cos\theta_2 = 45.4\%$$

.... (1.15)

It is now clear that only about 20% of the photons can be escaped from the devices, while the remaining photons are trapped inside the devices due to the TIR at the interface of glass/air and ITO/glass. To improve the outcoupling efficiency of the devices, a variety of methods have been proposed; for instance, texturing or roughening the substrate [64 - 70], patterning a microlens array [29, 30, 71, 72] or employing a photonic crystal structure [73, 74] were proposed to extract the substrate waveguide light; employing substrate with high refractive index [27, 28], inserting scattering media like silica spheres [67], TiO_2 nanoparticles [75] or low-index grids [76] between substrate and ITO were employed to extract the ITO/organic waveguide light. A maximum enhancement factor of 2.3 has been experimentally demonstrated by combining a high refractive index substrate or embedded low-index grids with a microlens array [28, 76].

FABRICATION AND CHARACTERIZATION OF OLEDS

Fig. (**1.9**) shows a typical device structure for the OLEDs, which consisted of several organic layers sandwiched between two electrodes.

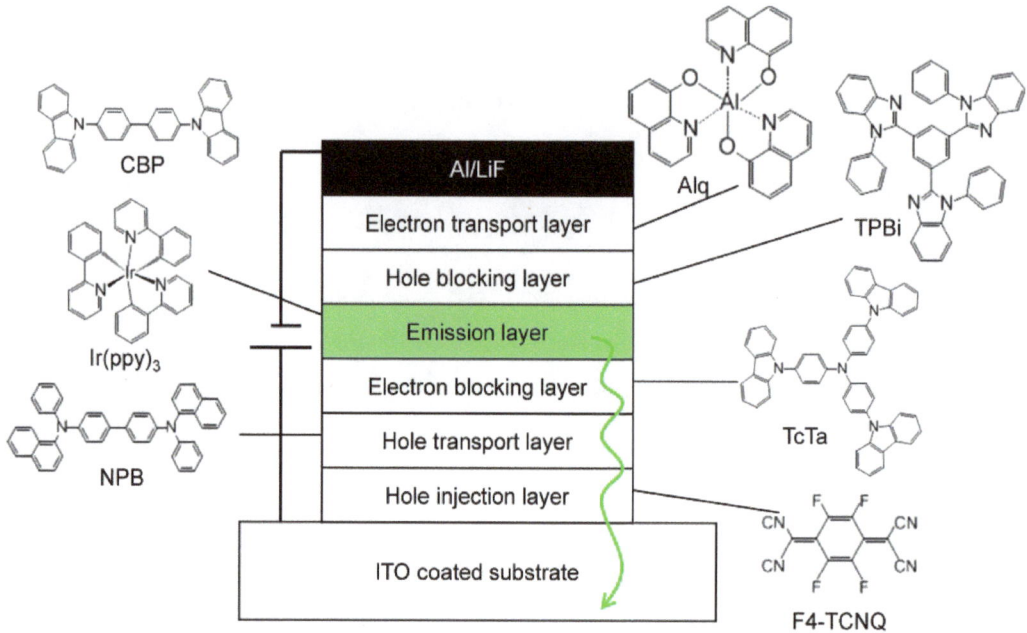

Fig. (1.9). Typical device structure of the bottom-emitting OLEDs.

Typical fabrication procedures for an OLED are shown in Fig. (**1.10**). The devices often are fabricated on a glass substrate that is pre-coated by ITO. The sheet resistance of the transparent ITO electrode typically ranges from 10 to 40 $\Omega/\square$ with a transmittance of around 80%-85%. The ITO is patterned using traditional lithography. Before depositing the organic layers, the ITO-coated glass substrates are thoroughly cleaned in ultrasonic acetone solution or detergent followed by spraying with deionized (DI) water and baking in an oven. The cleaned substrates are then treated by O_2 or CF_4 plasma to improve the work function of ITO [17, 18]. Organic layers like the hole-injection, hole-transport, electron-block, light emission, hole-block and electron-transport layers are then thermally sublimed in sequence using a vacuum evaporator at a base pressure of 10^{-6} to 10^{-7} Torr. The organic layers are patterned by the fine metal shadow masks. Ultra thin LiF layer capped with an Al layer are then deposited as a reflective cathode for the devices in a metal chamber [16]. The thickness of the deposited layers is monitored *in situ* by a quartz crystal microbalance. Doping of organic materials is realized by simultaneously co-evaporating the host and dopant materials. The deposition rate for the host materials typically is 1-2 angstrom/s, while for the dopant materials, according to the doping concentration, ranges from 0.01 to 0.1 angstrom/s. After finishing the deposition, the devices are then transferred to an Ar or N_2 filled glove box for encapsulation. A desiccant is attached with a glass or metal cap followed by assembling the protective cap with the devices using a UV-curable epoxy.

The current density (J)-voltage (V)-luminance (L) characteristics of the devices are then measured by a programmable source-measurement-unit (SMU) combined with a photodiode or a luminance meter. The external quantum efficiency, power efficiency and current efficiency can be inferred from the J-V-L curve, which is calculated by:

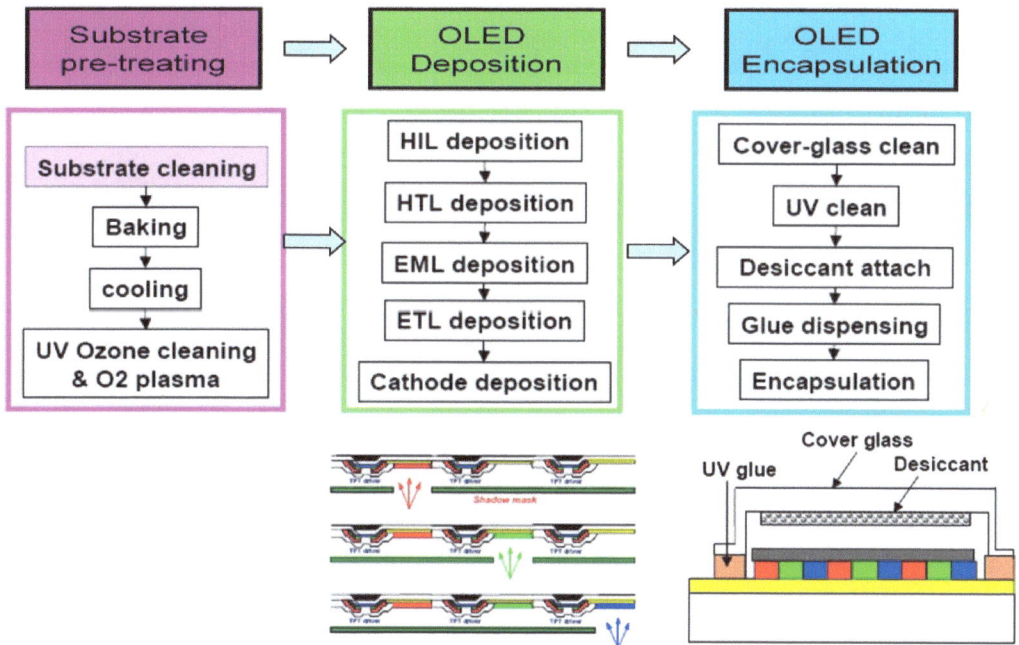

Fig. (1.10). Fabrication procedures for an OLED.

$$\eta_{current}\left(cd\middle/A\right) = \frac{L\left(cd\middle/m^2\right)}{J\left(A\middle/m^2\right)}$$ .... (1.16)

$$\eta_{power}\left(lm\middle/W\right) = \frac{\Omega(sr) \times L\left(cd\middle/m^2\right)}{J\left(A\middle/m^2\right) \times V(V)} = \frac{\Omega(sr) \times \eta_{current}\left(cd\middle/A\right)}{V(V)}$$ .... (1.17)

$$\eta_{EQE}(\%) = \frac{No.\,of\,\,photons}{No.\,of\,\,electrons} = \frac{\Omega(sr) \times L\left(cd\middle/m^2\right)\middle/ \dfrac{\int R(\lambda) \times S(\lambda) \times \dfrac{hc}{\lambda} d\lambda}{\int S(\lambda)\,d\lambda}}{J\left(A\middle/m^2\right)\middle/ e(C)} = \frac{eV\eta_{power}\left(lm\middle/W\right)}{\dfrac{\int R(\lambda) \times S(\lambda) \times \dfrac{hc}{\lambda} d\lambda}{\int S(\lambda)\,d\lambda}}$$ .... (1.18)

where Ω is the emission solid angle of the devices which depends on the emission angular distribution $I(\theta,\varphi)$ within the half-sphere in the forward direction. For Lambertian emission, $I(\theta,\varphi)=I_0\cos\theta$, thus [77]:

$$\Omega = \frac{1}{I_0}\int_0^{\frac{\pi}{2}}\int_{-\pi}^{\pi} I(\theta,\varphi)\sin\theta\,d\varphi d\theta = \frac{1}{I_0}\int_0^{\frac{\pi}{2}}\int_{-\pi}^{\pi} I_0\cos\theta\sin\theta\,d\varphi d\theta = \pi$$ .... (1.19)

$S(\lambda)$ are the emission spectra of the devices, which are measured by a spectrometer such as PR650. $R(\lambda)$ are the photopic response spectra of human eyes. As human eyes are most sensitive at the green light region, the luminance of the green emission is significantly higher than that of the blue or red emission at a certain power. Thus, the luminance, the current efficiency and the power efficiency are wavelength dependent, while the external quantum efficiency is wavelength irrelevant. For example, at an external quantum efficiency of 20%, the current efficiency of the Ir(ppy)$_3$ based green OLEDs (520 nm) is 70 cd/A, remarkably higher than 40 cd/A and 37 cd/A for the FirPic and Ir(2phq)$_2$(acac) based blue and red OLEDs, respectively.

The color of the emission is characterized by the CIE coordinates. Given the emission spectra $S(\lambda$, the tristimulus X, Y, Z can be calculated by:

$$X = \int_{380}^{780} S(\lambda)x(\lambda)d\lambda, \, Y = \int_{380}^{780} S(\lambda)y(\lambda)d\lambda, \, Z = \int_{380}^{780} S(\lambda)z(\lambda)d\lambda$$

$$\dots\dots\dots\dots (1.20)$$

where $x(\lambda), y(\lambda) \, and \, z(\lambda)$ are the color matching functions as shown in Fig. (**1.11**). The CIE coordinates are then calculated by:

$$x = \frac{X}{X+Y+Z}, \, y = \frac{Y}{X+Y+Z} \dots\dots\dots\dots\dots\dots\dots\dots\dots\dots (1.21)$$

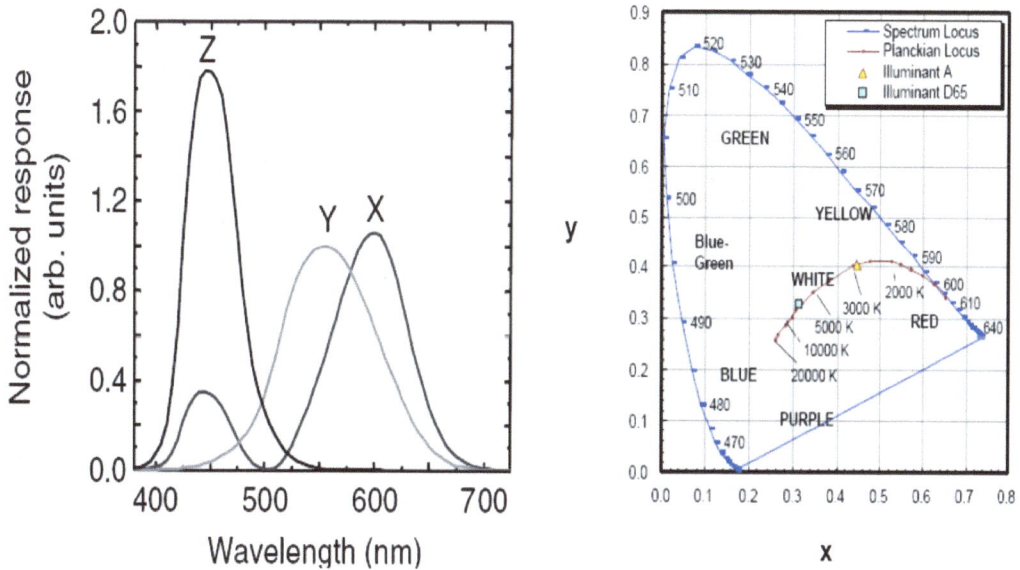

Fig. (1.11). Left: CIE 1931 color matching functions; right: CIE 1931 (x, y) chromaticity diagram [78].

Plots of the CIE coordinates of the monochromatic light give a horseshoe-shaped diagram as shown in Fig. (**1.11**). The CIE coordinate of any colors can be found in this diagram. An equal-energy white point with a CIE coordinates of (0.33, 0.33) represents a pure white color, which can be found in the center of the color diagram. The curve as plotted in the center of the diagram is the so-called Planckian locus. The colors on the Planckian locus are described by color temperature, which is the corresponding temperature at which the blackbody can emit the light of this color. The temperature for white color ranges from 2500 K (reddish warm white) to 20000 K (bluish cool white), which is located near the Planckian locus. The point labeled "Illuminant D65" and "Illuminant A" is the typical color of day light, and the color of an incandescent lamp, respectively, as standardized by the CIE. For the non-blackbody radiator having colors off from the Planckian locus, its chromaticity can be specified by the correlated color temperature (CCT), which refers to the temperature of a blackbody that has a color that is close to the emission from a non-blackbody radiator [79].

For lighting applications, the CIE coordinates are not enough to describe the quality of the light source. Although difference light sources can have the same CIE coordinates and exhibit the same color when viewed directly, their ability to render the color of an object under illumination may be significantly different. Therefore, to characterize the white light source, it is necessary to introduce the color rendition index (CRI), which is used to measure the ability of a light source to reproduce the colors of objects under illumination. The CRI is calculated by comparing light source to be tested with a reference source on the basis of how they render the eight sample colors. If the CCT of the light source to be tested is less than 5000 K, then a black body radiator (approximately like an incandescent lamp) is used as the reference source. For sources with higher CCT, the reference is a specifically defined spectrum of daylight. The calculation of CRI can be referred to "CIE 13.3-1995, Method of Measuring and Specifying Color Rendering Properties of Light Sources" [80, 81]. Broadband white spectrum covering the whole visible light region usually leads to a high CRI, as most test color can be rendered correctly under the illumination of such broadband white light source. For indoor lighting applications, the CRI usually should be higher than 80, while for visual inspection purposes such as surgery, photography and exhibition of museums, high CRI over 95 is desirable.

APPLICATION OF OLEDS

Flat Panel Display

OLEDs are considered to be the next generation flat-panel display technology due to its various attractive advantages like self-emitting, fast response time, thin and flexible, vivid color, high contrast and wide viewing angle *etc*. Back in 1997, Pioneer demonstrated the first OLED display for automotive audio. The early OLED products are based on a passive-matrix (PM) driven method which offers a low cost scheme for small area low resolution display applications such as MP3, PDA or cell phone. In PM driving scheme, with the scanning signals provided by external circuits, the OLEDs are lighted up row by row. In one frame time T, each row is scanned and turned on T/N time, where N is the number of the rows. Hence to achieve an average luminance of L, each pixel should emit a peak luminance of L*N. With the increase of panel resolution (N), the luminance of each pixel has to be tremendously boosted, resulting in a short operational lifetime. Therefore the PM driven method is not applicable for the high resolution displays.

To realize high resolution large area displays, an active-matrix (AM) driven method was proposed. In AM method, each OLED pixel is equipped with a switching thin-film transistor (TFT), a driving TFT and a capacitor, as shown in Fig. (**1.12**). When the switching TFT is turned on by the row scanning signal, the data current charges the capacitor and controls the driving TFT to turn on/off the OLED. After the scanning signal is gone, the data signal which is already stored in the capacitor continually controls the driving TFT to turn on/off the OLED. If the discharging time of the capacitor is significantly longer than the frame time, then the luminance of the OLED can be steadily maintained in the whole frame time T. In this way, the luminance of the OLED pixel is independent to the row number; thus large area and high resolution displays are possible.

In the early AMOLED product, the a-Si TFTs, which have been widely used in AMLCD, is adopted to drive the OLEDs due to their mature and relatively simple fabrication process, cheap manufacturing cost and low capital investment [82]. However, in contrast with LCD, the OLED is a current driven device which means the a-Si TFTs are not a good partner for OLED, as they cannot provide sufficient current for driving the OLED due to their low mobility of ~1 cm^2/Vs [83].

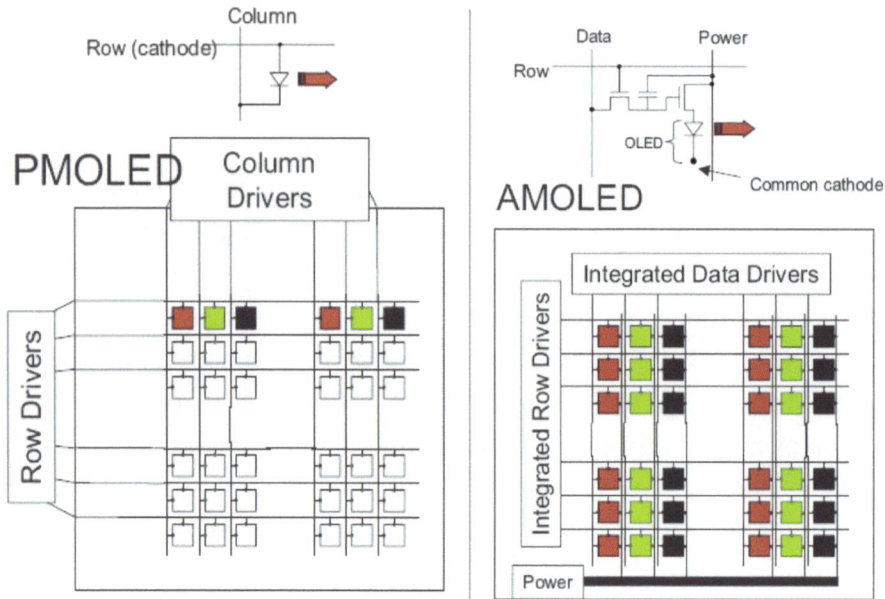

Fig. (1.12). Driving methods of PM- and AMOLED.

Low temperature poly silicon (LTPS) TFTs with higher mobility (~50 cm^2/Vs) are adopted for high resolution AMOLED [84]. However, the uniformity of the LTPS TFTs is poor mainly due to the grain boundary presence in the devices, making them problematic and costly for large area displays. Recently, ZnO (zinc oxide) or IGZO (indium gallium zinc oxide) based metal oxide TFTs have become an active and hot research field due to their high mobility (10 cm^2/Vs), transparency, simple and low fabrication cost - all rendering them a potentially ideal candidate for large area AMOLED [85, 86].

In 1999, Kodak & Sanyo developed the first full color LTPS AMOLED for camera display application. Since then, various AMOLED prototypes featuring a large area (55 inch), flexibility and transparency have been demonstrated, but there are still few products in the market. In 2007, Sony launched the first commercial 11 inch full color AMOLED TV XEL-1 with a high price of US$ 1740 [87]. Very recently, Samsung released a 4.65 inch high definition full color

AMOLED panel with resolution 1280 by 720 for their Galaxy Nexus mobile phone [88]. These two products are shown in Fig. (**1.13**).

Fig. (1.13). Left: Sony's commercialized 11 inch full color AMOLED TV XEL-1; Right: Samsung's 4.65 inch high definition full color AMOLED panel for Galaxy Nexus.

Although OLED exhibits lots of advantages compared with LCD as shown in Table **1**, it is currently challenging for them to dominate the market due to their smaller display area, relatively lower resolution, shorter lifetime and higher production cost than LCD. With emerging oxide TFT technology, new color patterning scheme, new blue stable emitters and improved hermetic sealing, large size, high resolution, long lifetime and low cost will potentially be achieved in the near future.

Table 1.1. Comparisons of AMOLED and AMLCD.

	AM OLED	AM LCD
Thickness/Weight	Thinner, best is 0.05 nm; lighter	Thicker, best is 0.3 nm; heavier
Diagonal Aize	Limited to small and medium size; largest demo is around 40"	Can be manufactured larger; largest demo is ~100"
Viewing Angle	Up to 180 degrees	Narrower, depends on liqiud crystal type
Color Gaamut	>100% NTSC (top emission), around 70% NTSC (bottom); high at all gray levels	Around 70% up to 100% NTSC (LED backlight and new color filter); falls at low gray levels
Color Reproduction	Better; gamut independent of view angle	Good gamut changes with viewing angle

(Table 1.1) contd.....

	AM OLED	AM LCD
Resolution	Lower; 308 dpi (SM), 202 dpi (polymer)	Higher; best is 498 dpi
Response Time	Faster, nanoseconds, No motion blur, good for 3D.	Slower, milliseconds
Contrast Ratio	Higher	Lower
Sunlight Readability	Better than transmissive LCD, worse than transflective LED.	OK if transflective
Operating Temperature	Range is large, can operate at low temps like -40 °C	Range is smaller, lowest temp is -10 °C
Power Consumption	Lower at typical video content, whem around 30% of pixels are on.	Higher at typical video content
lifetime	Shorter, 5K to 30K hour, but improving	Much longer, above 50K hour.
Manufacturing Investment	lower, but lack of standards keeps the investment only slightly lower	Higher
Production Cost	Expensive, low yield and complex structure, petential to be low cost.	Cheap than AMOLED

Based on the forecast by DisplaySearch [89], the OLED display revenues will grow to about US$ 8B in 2017, up from US$ 0.8B in 2009, with major applications for cell phone display and TV, as shown in Fig. (**1-14**).

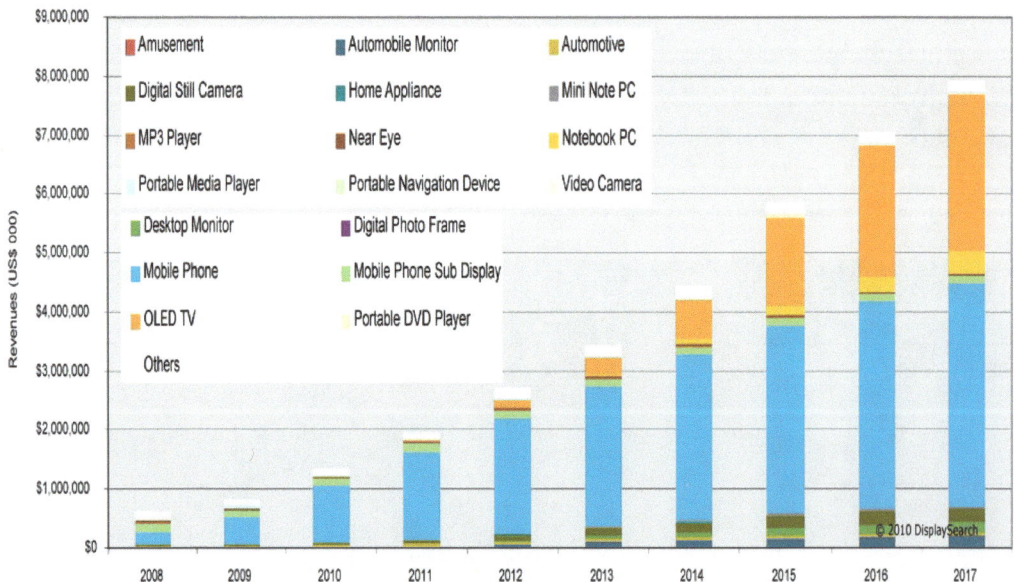

Fig. (1.14). OLED long-term annual revenue forecast [89].

Solid-state Lighting

Another important application of OLEDs is solid-state lighting. As lighting sources, OLEDs have many inherent advantages. For instance, at a illumination level of 1000 cd/m^2, the white OLEDs can have a high efficiency of 100 lm/W [90], which is comparable with that of the fluorescent tube and beyond that of the incandescent bulb. Furthermore, heat dissipation is not as demanding as that of LED because they are true surface/area lighting sources. Also, the color is tunable for color matching or decorative purposes. It is environmentally friendly because it does not contain toxic mercury. More importantly, it can be transparent like glass, flexible like cloth and reflective like mirror, which will open new design and architectural opportunities; for example, it allows the designers to integrate the lighting with the window (Fig. **1.15**) which could not be possible without the aid of transparent OLEDs. Many companies such as OSRAM, Philips, Lumiotec, GE and Konica Minolta have participated actively in the development of OLED lighting. According to the forecast by DisplaySearch (shown in Fig. (**1.16**)), the OLED lighting market will reach US$1.5 B by 2015 and US$ 6.3B by 2018 [91].

Fig. (1.15). OLED lighting: new concept for new lifestyle.

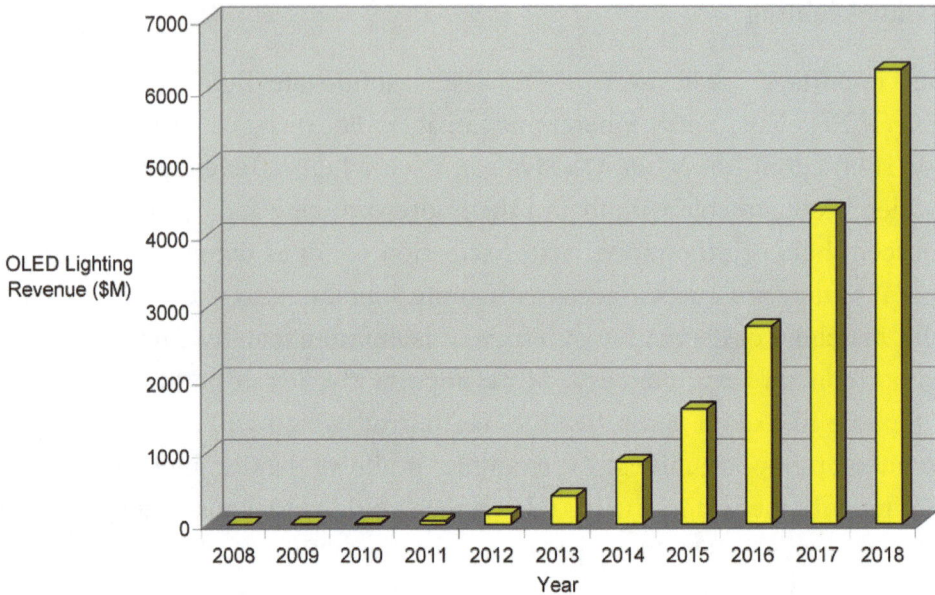

Fig. (1.16). OLED lighting market forecast [91].

CONFLICT OF INTEREST

The authors confirm that this chapter contents have no conflict of interest.

ACKNOWLEDGMENTS

This work was supported by the National Natural Science Foundation of China under Grant No. 61405089, and by the Innovation of Science and Technology Committee of Shenzhen under Grant No. JCYJ20140417105742713.

REFERENCES

[1]   C.W. Tang, and S.A. VanSlyke, "Organic electroluminescent diodes", *Appl. Phys. Lett.,* vol. 51, pp. 913-915, 1987.
        [http://dx.doi.org/10.1063/1.98799]

[2]   C.W. Tang, S.A. VanSlyke, and C.H. Chen, "Electroluminescence of doped organic thin films", *J. Appl. Phys.,* vol. 65, pp. 3610-3616, 1989.
        [http://dx.doi.org/10.1063/1.343409]

[3]   L.S. Hung, and C.H. Chen, "Recent progress of molecular organic electroluminescent materials and devices", *Mater. Sci. Eng. Rep.,* vol. 39, pp. 143-222, 2002.
        [http://dx.doi.org/10.1016/S0927-796X(02)00093-1]

[4]   C-C. Wu, C-W. Chen, C-L. Lin, and C-J. Yang, "Advanced organic light-emitting devices for

enhancing display performances", *J. Disp. Technol.,* vol. 01, pp. 248-266, 2005.
[http://dx.doi.org/10.1109/JDT.2005.858942]

[5]   C.W. Tang, "Twolayer organic photovoltaic cell", *Appl. Phys. Lett.,* vol. 48, pp. 183-185, 1986.
[http://dx.doi.org/10.1063/1.96937]

[6]   J. Xue, S. Uchida, B.P. Rand, and S.R. Forrest, "4.2% efficient organic photovoltaic cells with low
series resistances", *Appl. Phys. Lett.,* vol. 84, pp. 3013-3015, 2004.
[http://dx.doi.org/10.1063/1.1713036]

[7]   P. Peumans, A. Yakimov, and S.R. Forrest, "Small molecular weight organic thin-film photodetectors
and solar cells", *J. Appl. Phys.,* vol. 93, pp. 3693-3723, 2003.
[http://dx.doi.org/10.1063/1.1534621]

[8]   H.K. Tsumura, and T. Ando, "Macromolecular electronic device: field effect transistor with a
polythiophene thin film", *Appl. Phys. Lett.,* vol. 49, pp. 1210-1212, 1986.
[http://dx.doi.org/10.1063/1.97417]

[9]   C.D. Dimitrakopoulos, and D.J. Mascaro, "Organic thin-film transistors: a review of recent advances",
IBM J. Res. Develop., vol. 45, pp. 11-27, 2001.
[http://dx.doi.org/10.1147/rd.451.0011]

[10]   M. Pope, H. Kallman, and P. Magnante, "Electroluminescence in organic crystals", *J. Chem. Phys.,*
vol. 38, p. 2042, 1963.
[http://dx.doi.org/10.1063/1.1733929]

[11]   W. Helfrich, and W.G. Schneider, "Recombination radiation in anthracene crystals", *Phys. Rev. Lett.,*
vol. 14, p. 229, 1965.
[http://dx.doi.org/10.1103/PhysRevLett.14.229]

[12]   J.H. Burroughes, D.D. Bradley, A.R. Brown, R.N. Marks, K. Mackay, R.H. Friend, P.L. Burn, and
A.B. Holmes, "Light-emitting diodes based on conjugated polymers", *Nature,* vol. 347, p. 539, 1990.
[http://dx.doi.org/10.1038/347539a0]

[13]   J. Shi, and C.W. Tang, "Doped organic electroluminescent devices with improved stability", *Appl.
Phys. Lett.,* vol. 70, p. 13, 1997.
[http://dx.doi.org/10.1063/1.118664]

[14]   G. Sakamoto, C. Adachi, T. Koyama, and Y. Taniguchi, "Significant improvement of device durability
in organic light-emitting diodes by doping both hole transport and emitter layers with rubrene
molecules", *Appl. Phys. Lett.,* vol. 75, p. 766, 1999.
[http://dx.doi.org/10.1063/1.124506]

[15]   C. Adachi, T. Tsutsui, and S. Saito, "Confinement of charge carriers and molecular excitons within 5-
nm-thick emitter layer in organic electroluminescent devices with a double heterostructure", *Appl.
Phys. Lett.,* vol. 57, p. 531, 1990.
[http://dx.doi.org/10.1063/1.103638]

[16]   L.S. Hung, C.W. Tang, and M.G. Mason, "Enhanced electron injection in organic electroluminescence devices using an Al/LiF electrode", *Appl. Phys. Lett.,* vol. 70, p. 152, 1997.
[http://dx.doi.org/10.1063/1.118344]

[17]   C.C. Wu, C.I. Wu, J.C. Sturm, and K. Kahn, "Surface modification of indium tin oxide by plasma

treatment: An effective method to improve the efficiency, brightness, and reliability of organic light emitting devices", *Appl. Phys. Lett.,* vol. 70, p. 1348, 1997.
[http://dx.doi.org/10.1063/1.118575]

[18] L.S. Huang, L.R. Zheng, and M.G. Mason, "Anode modification in organic light-emitting diodes by low-frequency plasma polymerization of CHF3", *Appl. Phys. Lett.,* vol. 78, p. 673, 2001.
[http://dx.doi.org/10.1063/1.1331639]

[19] M.A. Baldo, D.F. O'Brien, Y. You, A. Shoustikov, S. Sibley, M.E. Thompson, and S.R. Forrest, "Highly efficient phosphorescent emission from organic electroluminescent devices", *Nature,* vol. 395, p. 151, 1998.
[http://dx.doi.org/10.1038/25954]

[20] M.A. Baldo, S. Lamansky, P.E. Burrows, M.E. Thompson, and S.R. Forrest, "Very high-efficiency green organic light-emitting devices based on electrophosphorescence", *Appl. Phys. Lett.,* vol. 75, p. 4, 1999.
[http://dx.doi.org/10.1063/1.124258]

[21] C. Adachi, M.A. Baldo, M.E. Thompson, and S.R. Forrest, "Nearly 100% internal phosphorescence efficiency in an organic light-emitting device", *J. Appl. Phys.,* vol. 90, p. 5048, 2001.
[http://dx.doi.org/10.1063/1.1409582]

[22] J. Kido, and T. Matsumoto, "Bright organic electroluminescent devices having a metal-doped electron-injecting layer", *Appl. Phys. Lett.,* vol. 73, p. 2866, 1998.
[http://dx.doi.org/10.1063/1.122612]

[23] X. Zhou, M. Pfeiffer, J. Blochwitz, A. Nollau, T. Fritz, and K. Leo, "Very-low-operating-voltage organic light-emitting diodes using a p-doped amorphous hole injection layer", *Appl. Phys. Lett.,* vol. 78, p. 410, 2001.
[http://dx.doi.org/10.1063/1.1343849]

[24] K. Walzer, B. Maennig, M. Pfeiffer, and K. Leo, "Highly efficient organic devices based on electrically doped transport layers", *Chem. Rev.,* vol. 107, no. 4, pp. 1233-1271, 2007.
[http://dx.doi.org/10.1021/cr050156n] [PMID: 17385929]

[25] J. Huang, M. Pfeiffer, A. Werner, J. Blochwitz, K. Leo, and S. Liu, "Low-voltage organic electroluminescent devices using pin structures", *Appl. Phys. Lett.,* vol. 80, p. 139, 2002.
[http://dx.doi.org/10.1063/1.1432110]

[26] K. Saxena, V.K. Jain, and D.S. Mehta, "A review on the light extraction techniques in organic electroluminescent devices", *Opt. Mater.,* vol. 32, pp. 221-233, 2009.
[http://dx.doi.org/10.1016/j.optmat.2009.07.014]

[27] S. Reineke, F. Lindner, G. Schwartz, N. Seidler, K. Walzer, B. Lüssem, and K. Leo, "White organic light-emitting diodes with fluorescent tube efficiency", *Nature,* vol. 459, no. 7244, pp. 234-238, 2009.
[http://dx.doi.org/10.1038/nature08003] [PMID: 19444212]

[28] A. Mikami, and T. Koyanagi, "60.4L: late-news paper: high efficiency 200-lm/W green light emitting organic devices prepared on high-index of refraction substrates", *SID Symp. Dig. Tech. Papers,* vol. 40, pp. 907-910, 2009.
[http://dx.doi.org/10.1889/1.3256944]

[29] S. Möller, and S.R. Forrest, "Improved light out-coupling in organic light emitting diodes employing ordered microlens arrays", *J. Appl. Phys.*, vol. 91, pp. 3324-3327, 2002.
[http://dx.doi.org/10.1063/1.1435422]

[30] H. Peng, Y.L. Ho, X-J. Yu, M. Wong, and H-S. Kwok, "Coupling efficiency enhancement in organic light-emitting devices using microlens array-theory and experiment", *J. Disp. Technol.*, vol. 01, pp. 278-282, 2005.
[http://dx.doi.org/10.1109/JDT.2005.858944]

[31] W. Brütting, S. Berrleb, and A.G. Mückl, "Device physics of organic light-emitting diodes based on molecular materials", *Org. Electron.*, vol. 02, pp. 1-36, 2001.
[http://dx.doi.org/10.1016/S1566-1199(01)00009-X]

[32] S.M. Sze, *Physics of semiconductor devices.* Wiley: New York, 1981.

[33] X. Zhu, J. Sun, X. Yu, M. Wong, and H-S. Kwok, "Investigation of Al- and Ag-based top-emitting organic light-emitting diodes with metal oxides as hole-injection layer", *Jpn. J. Appl. Phys.*, vol. 46, pp. 1033-1036, 2007.
[http://dx.doi.org/10.1143/JJAP.46.1033]

[34] M.Y. Chan, S.L. Lai, M.K. Fung, C.S. Lee, and S.T. Lee, "Impact of the metal cathode and CsF buffer layer on the performance of organic light-emitting devices", *J. Appl. Phys.*, vol. 95, p. 5397, 2004.
[http://dx.doi.org/10.1063/1.1707201]

[35] J. Lee, Y. Park, D.Y. Kim, H.Y. Chu, H. Lee, and L-M. Do, "High efficiency organic light-emitting devices with Al/NaF cathode", *Appl. Phys. Lett.*, vol. 82, p. 173, 2003.
[http://dx.doi.org/10.1063/1.1537048]

[36] Q. Xu, J. Ouyang, Y. Yang, T. Ito, and J. Kido, "Ultrahigh efficiency green polymer light-emitting diodes by nanoscale interface modification", *Appl. Phys. Lett.*, vol. 83, p. 4695, 2003.
[http://dx.doi.org/10.1063/1.1630848]

[37] Y-H. Niu, Q. Hou, and Y. Cao, "Thermal annealing below the glass transition temperature: a general way to increase performance of light-emitting diodes based on copolyfluorenes", *Appl. Phys. Lett.*, vol. 81, p. 634, 2002.
[http://dx.doi.org/10.1063/1.1495898]

[38] G.Z. Ran, G.L. Ma, Y.H. Xu, L. Dai, and G.G. Qin, "Light extraction efficiency of a top-emission organic light-emitting diode with an Yb/Au double-layer cathode and an opaque Si anode", *Appl. Opt.*, vol. 45, no. 23, pp. 5871-5876, 2006.
[http://dx.doi.org/10.1364/AO.45.005871] [PMID: 16926874]

[39] T. Oyamada, H. Sasabe, C. Adachi, S. Murase, T. Tominaga, and C. Maeda, "Extremely low-voltage driving of organic light-emitting diodes with a Cs-doped phenyldipyrenylphosphine oxide layer as an electron-injection layer", *Appl. Phys. Lett.*, vol. 86, p. 033503, 2005.
[http://dx.doi.org/10.1063/1.1852707]

[40] P-C. Kao, J-H. Lin, J-Y. Wang, C-H. Yang, and S-H. Chen, "Li2CO3 as an n-type dopant on Alq3-based organic light emitting devices", *J. Appl. Phys.*, vol. 109, p. 094505, 2011.
[http://dx.doi.org/10.1063/1.3585767]

[41] T. Xiong, F. Wang, X. Qiao, and D. Ma, "Cesium hydroxide doped tris-(8-hydroxyquinoline)

aluminum as an effective electron injection layer in inverted bottom-emission organic light emitting diodes", *Appl. Phys. Lett.,* vol. 92, p. 263305, 2008.
[http://dx.doi.org/10.1063/1.2955516]

[42] S.K. So, W.K. Choi, C.H. Cheng, L.M. Leung, and C.F. Kwong, "Surface preparation and characterization of indiumtin oxide substrates for organic electroluminescent devices", *Appl. Phys., A Mater. Sci. Process.,* vol. 68, pp. 447-450, 1999.
[http://dx.doi.org/10.1007/s003390050921]

[43] J. Li, M. Yahiro, K. Ishida, H. Yamada, and K. Matsushige, "Enhanced performance of organic light emitting device by insertion of conducting/insulating WO3 anodic buffer layer", *Synth. Met.,* vol. 151, pp. 141-146, 2005.
[http://dx.doi.org/10.1016/j.synthmet.2005.03.019]

[44] J. Meyer, T. Winkler, S. Hamwi, S. Schmale, H-H. Johannes, T. Weimann, P. Hinze, W. Kowlasky, and T. Riedl, "Transparent inverted organic light-emitting diodes with a tungsten oxide buffer layer", *Adv. Mater.,* vol. 20, pp. 3839-3843, 2008.
[http://dx.doi.org/10.1002/adma.200800949]

[45] H. Lee, S.W. Cho, K. Han, P.E. Jeon, C-N. Whang, K. Jeong, K. Cho, and Y. Yi, "The origin of the hole injection improvements at indium tin oxide/molybdenum trioxide/N,N-bis(1-naphthyl)- N,N-diphenyl-1,1-biphenyl- 4,4-diamine interfaces", *Appl. Phys. Lett.,* vol. 93, p. 043308, 2008.
[http://dx.doi.org/10.1063/1.2965120]

[46] H. You, Y. Dai, Z. Zhang, and D. Ma, "Improved performances of organic light-emitting diodes with metal oxide as anode buffer", *J. Appl. Phys.,* vol. 101, p. 026105, 2007.
[http://dx.doi.org/10.1063/1.2430511]

[47] T. Matsushima, G-H. Jin, and H. Murata, "Marked improvement in electroluminescence characteristics of organic light-emitting diodes using an ultrathin hole-injection layer of molybdenum oxide", *J. Appl. Phys.,* vol. 104, p. 054501, 2008.
[http://dx.doi.org/10.1063/1.2974089]

[48] X-Y. Jiang, Z-L. Zhang, J. Cao, M.A. Khan, K. Haq, and W-Q. Zhu, "White OLED with high stability and low driving voltage based on a novel bufferlayer MoOx", *J. Phys. D Appl. Phys.,* vol. 40, pp. 5553-5557, 2007.
[http://dx.doi.org/10.1088/0022-3727/40/18/007]

[49] H.M. Zhang, and W.C. Choy, "Highly efficient organic light-emitting devices with surface-modified metal anode by vanadium pentoxide", *J. Phys. D Appl. Phys.,* vol. 41, p. 062003, 2008.
[http://dx.doi.org/10.1088/0022-3727/41/6/062003]

[50] J. Wu, J. Hou, Y. Cheng, Z. Xie, and L. Wang, "Efficient top-emitting organic light-emitting diodes with a V2O5 modified silver anode", *Semicond. Sci. Technol.,* vol. 22, pp. 824-826, 2007.
[http://dx.doi.org/10.1088/0268-1242/22/7/027]

[51] G. Xie, Y. Meng, F. Wu, C. Tao, D. Zhang, M. Liu, Q. Xue, W. Chen, and Y. Zhao, "Very low turn-on voltage and high brightness tris-8-hydroxyquinoline aluminum-based organic light-emitting diodes with a MoOx p-doping layer", *Appl. Phys. Lett.,* vol. 92, p. 093305, 2008.
[http://dx.doi.org/10.1063/1.2890490]

[52] W-J. Shin, J-Y. Lee, J.C. Kim, T-H. Yoon, T-S. Kim, and O-K. Song, "Bulk and interface properties

of molybdenum trioxide-doped hole transporting layer in organic light-emitting diodes", *Org. Electron.,* vol. 09, pp. 333-338, 2008.
[http://dx.doi.org/10.1016/j.orgel.2007.12.001]

[53] J. Qiu, Z.B. Wang, M.G. Helander, and Z.H. Lu, "MoO3 doped 4,4-N,N-dicarbazole-biphenyl for low voltage organic light emitting diodes", *Appl. Phys. Lett.,* vol. 99, p. 153305, 2011.
[http://dx.doi.org/10.1063/1.3644159]

[54] J. Meyer, S. Hamwi, S. Schmale, T. Winkler, H-H. Johannes, T. Riedl, and W. Kowalsky, "A strategy towards p-type doping of organic materials with HOMO levels beyond 6 eV using tungsten oxide", *J. Mater. Chem.,* vol. 19, pp. 702-705, 2009.
[http://dx.doi.org/10.1039/b819485h]

[55] F. Wang, X. Qiao, T. Xiong, and D. Ma, "Tungsten oxide doped N,N-di(naphthalen-1-yl)-N,N-diphenylbenzidine as hole injection layer for high performance organic light emitting diodes", *J. Appl. Phys.,* vol. 105, p. 084518, 2009.
[http://dx.doi.org/10.1063/1.3116204]

[56] P.N. Murgatroyd, "Theory of space-charge-limited current enhanced by Frenkel effect", *J. Phys. D Appl. Phys.,* vol. 03, p. 151, 1970.
[http://dx.doi.org/10.1088/0022-3727/3/2/308]

[57] N.F. Mott, and R.W. Gurney, *Electronic Processes in Ionic Crystals.* Clarendon Press: Oxford, 1940.

[58] P.E. Burrows, Z.L. Shen, V. Bulovic, D.M. Mccarty, and S.R. Forrest, "Relationship between electroluminescence and current transport in organic heterojunction light-emitting devices", *J. Appl. Phys.,* vol. 79, p. 7991, 1996.
[http://dx.doi.org/10.1063/1.362350]

[59] M.A. Lampert, and P. Mark, *Current Injection in Solids.* Academic Press: New York, 1970.

[60] J. Staudigel, M. Stößel, F. Steuber, and J. Simmerer, "A quantitative numerical model of multilayer vapor-deposited organic light emitting diodes", *J. Appl. Phys.,* vol. 86, p. 3895, 1999.
[http://dx.doi.org/10.1063/1.371306]

[61] D. Ruhstaller, S.A. Carter, S. Barth, H. Riel, W. Riess, and J.C. Scott, "Transient and steady-state behavior of space charges in multilayer organic light-emitting diodes", *J. Appl. Phys.,* vol. 89, p. 4575, 2001.
[http://dx.doi.org/10.1063/1.1352027]

[62] Z. Kafafi, *Organic Electroluminescene.* CRC Press: New York, 2005.

[63] R.J. Silbery, R.A. Alberty, and M.G. Bawendi, *Physical Chemistry.* Wiley: Hoboken, NJ, 2005.

[64] T. Nakamura, and N. Tsutsumi, "Improvement of coupling-out efficiency in organic electroluminescent devices by addition of a diffusive layer", *J. Appl. Phys.,* vol. 96, p. 6016, 2004.
[http://dx.doi.org/10.1063/1.1810196]

[65] S. Chen, and H-S. Kwok, "Light extraction from organic light-emitting diodes for lighting applications by sand-blasting substrates", *Opt. Express,* vol. 18, no. 1, pp. 37-42, 2010.
[http://dx.doi.org/10.1364/OE.18.000037] [PMID: 20173819]

[66] R. Bathelt, D. Buchhauser, C. Gärditz, R. Paetzold, and P. Wellmann, "Light extraction from OLEDs for lighting applications through light scattering", *Org. Electron.,* vol. 08, pp. 293-299, 2007.

[http://dx.doi.org/10.1016/j.orgel.2006.11.003]

[67]   T. Yamasaki, K. Sumioka, and T. Tsutsui, "Organic light-emitting device with an ordered monolayer of silica microspheres as a scattering medium", *Appl. Phys. Lett.,* vol. 76, pp. 1243-1245, 2000. [http://dx.doi.org/10.1063/1.125997]

[68]   Y-H. Cheng, J-L. Wu, C-H. Cheng, K-C. Syao, and M-C. Lee, "Enhanced light outcoupling in a thin film by texturing meshed surfaces", *Appl. Phys. Lett.,* vol. 90, p. 091102, 2007. [http://dx.doi.org/10.1063/1.2709920]

[69]   J.J. Shiang, T.J. Faircloth, and A.R. Duggal, "Experimental demonstration of increased organic light emitting device output via volumetric light scattering", *J. Appl. Phys.,* vol. 95, pp. 2889-2895, 2004. [http://dx.doi.org/10.1063/1.1644038]

[70]   Y-H. Ho, C-C. Liu, S-W. Liu, H. Liang, C-W. Chu, and P-K. Wei, "Efficiency enhancement of flexible organic light-emitting devices by using antireflection nanopillars", *Opt. Express,* vol. 19, suppl. Suppl. 3, pp. A295-A302, 2011. [http://dx.doi.org/10.1364/OE.19.00A295] [PMID: 21643370]

[71]   J.Y. Kim, and K.C. Choi, "Improvement in Outcoupling Efficiency and Image Blur of Organic Light-Emitting Diodes by Using Imprinted Microlens Arrays", *J. Disp. Technol.,* vol. 07, pp. 377-381, 2011. [http://dx.doi.org/10.1109/JDT.2011.2121053]

[72]   J-H. Lee, Y-H. Ho, K-Y. Chen, H-Y. Lin, J-H. Fang, S-C. Hsu, J-R. Lin, and M-K. Wei, "Efficiency improvement and image quality of organic light-emitting display by attaching cylindrical microlens arrays", *Opt. Express,* vol. 16, no. 26, pp. 21184-21190, 2008. [http://dx.doi.org/10.1364/OE.16.021184] [PMID: 19104547]

[73]   Y.J. Lee, S.H. Kim, J. Hun, G.H. Kim, Y.H. Lee, S.H. Cho, Y.C. Kim, and Y.R. Do, "A high-extraction-efficiency nanopatterned organic light-emitting diode", *Appl. Phys. Lett.,* vol. 82, pp. 3779-3781, 2003. [http://dx.doi.org/10.1063/1.1577823]

[74]   K. Ishihara, M. Fujita, I. Matsubara, T. Asano, S. Noda, H. Ohata, A. Hirasawa, H. Nakada, and N. Shimoji, "Organic light-emitting diodes with photonic crystals on glass substrate fabricated by nanoimprint lithography", *Appl. Phys. Lett.,* vol. 90, p. 111114, 2007. [http://dx.doi.org/10.1063/1.2713237]

[75]   H-W. Chang, K-C. Tien, M-H. Hsu, Y-H. Huang, M-S. Lin, C-H. Tsai, Y.T. Tsai, and C-C. Wu, "Organic light-emitting devices integrated with internal scattering layers for enhancing optical out-coupling", *J. Soc. Inf. Disp.,* vol. 19, pp. 196-204, 2011. [http://dx.doi.org/10.1889/JSID19.2.196]

[76]   Y. Sun, and S.R. Forrest, "Enhanced light out-coupling of organic light-emitting devices using embedded low-index grids", *Nat. Photonics,* vol. 02, pp. 483-487, 2008. [http://dx.doi.org/10.1038/nphoton.2008.132]

[77]   G. Schwartz, S. Reineke, T.C. Rosenow, K. Walzer, and K. Keo, "Triplet Harvesting in Hybrid White Organic Light-Emitting Diodes", *Adv. Funct. Mater.,* vol. 19, pp. 1319-1333, 2009. [http://dx.doi.org/10.1002/adfm.200801503]

[78]   Commission Internationale de L'éclairage (CIE) Colorimetry, *Publication Report No. 15.2,* 1986.

[79] Á. Borbély, Á. Sámson, and J. Schanda, "The concept of correlated color temperature revisited", *Color Res. Appl.,* vol. 26, pp. 450-457, 2001.
[http://dx.doi.org/10.1002/col.1065]

[80] Commission Internationale de L'éclairage (CIE), *Method of measuring and specifying color rendering properties of light sources, Publication Report No. 13.2,* 1974.

[81] Joint ISO/CIE Standard, *CIE standard illuminants for colorimetry provide explanations and descriptions of the CIE standard illuminants.* ISO 10526:1999/CIE S005/E-1998.

[82] P. Servati, S. Prakash, A. Nathan, and C. Py, "Amorphous silicon driver circuits for organic light-emitting diode displays", *J. Vac. Sci. Techo.,* vol. A 20, p. 1374, 2002.
[http://dx.doi.org/10.1116/1.1486006]

[83] J-J. Lih, C-F. Sung, C-H. Li, T-H. Hsiao, and H-H. Lee, "Comparison of a-Si and Poly-Si for AMOLED displays", *J. Soc. Inf. Disp.,* vol. 12, p. 367, 2004.
[http://dx.doi.org/10.1889/1.1847734]

[84] H-K. Chung, K-Y. Lee, and S.T. Lee, "Alternative approach to large-sized AMOLED HDTV", *J. Soc. Inf. Disp.,* vol. 14, p. 49, 2006.
[http://dx.doi.org/10.1889/1.2166835]

[85] H-H. Hsieh, T-T. Tsai, C-Y. Chang, H-H. Wang, J-Y. Huang, and Y-H. Lin, "11.2: A 2.4-in. AMOLED with IGZO TFTs and Inverted OLED Devices", *SID Symp. Dig. Tech. Papers,* vol. 41, pp. 140-143, 2010.
[http://dx.doi.org/10.1889/1.3499968]

[86] S-H. Park, M. Ryu, S. Yang, C. Byun, C-S. Hwang, and K.I. Cho, "18.1: Invited paper: Oxide TFT driving transparent AM-OLED", *SID Symp. Dig. Tech. Papers,* vol. 41, pp. 245-248, 2010.
[http://dx.doi.org/10.1889/1.3500418]

[87] http://www.oled-display.net/sony-xel-1-oled-tv/.

[88] http://www.theverge.com/2011/11/17/2568348/galaxy-nexus-review.

[89] http://www.displaysearch.com/cps/rde/xchg/displaysearch/ hs.xsl/quarterly_oled_shipment_and_forecast_report.asp.

[90] http://www.universaldisplay.com/downloads/Press%20Releases/2008/PANL_white-milestone_FINAL .pdf.

[91] http://www.displaysearch.com/cps/rde/xchg/displaysearch/hs.xsl/topical_oled_lighting_in_2009_ and_beyond_report.asp.

CHAPTER 2

White Organic Light-Emitting Diodes for Display and Lighting Application

Shuming Chen[*]

Department of Electrical and Electronic Engineering, South University of Science and Technology of China, Shenzhen, 518055, P. R. China

Abstract: The history of white light organic light-emitting diodes (WOLEDs) can be dated back to 1993 when Kido and coworkers reported the first OLEDs that emit white color, *i.e.*, the emitted light contain wavelengths across the entire visible spectrum [1, 2]. From then on, considerable research efforts have been devoted throughout the world by both academia and industry in developing efficient WOLEDs for next generation high resolution large area displays and solid-state light sources. In this chapter, the applications of WOLEDs in display and lighting will first be presented, followed by introducing the general approaches to achieve white light emission. Finally, challenges and solutions for large area, high efficiency and high CRI WOLEDs will be posed.

Keywords: Color patterning, Color rendering index, Emitting diode (WOLED), Solid-state lighting, White organic light- full color display.

WOLEDs FOR FULL COLOR DISPLAYS

In order to realize a full color display, the OLEDs should emit the red (R), green (G) and blue (B) three primary colors. A simple and straight forward method is to deposit the R, G and B emission layers side by side, as shown in Fig. (**2.1**). The emission layers are patterned by the fine metal shadow masks (FMMs). Organic functional materials are deposited and defined onto the substrate through the openings of the FMMs. The FMMs, with opening as small as 50 μm, typically are made by etching or electroforming the metal sheet [3]. Although FMMs is widely used in the production of full color OLED display currently, they are unsuitable for future high resolution and large area display, because they have some intrinsic

[*] **Corresponding Author Shuming Chen:** Department of Electrical and Electronic Engineering, South University of Science and Technology of China, Shenzhen, 518055, P. R. China; Tel.: +86-755-88018522; E-mail: chen.sm@sustc.edu.cn

limitations. For instance, the stainless steel used in FMMs always suffers from shape changing due to heat and/or external force, and when the size becomes larger it is hard to handle [3 - 5]. To circumvent these problems, alternative color patterning methods were proposed such as inkjet printing [6, 7], laser-induced thermal imaging and radiation induced sublimation transferring [8, 9]; however, all these new methods are still under developing and do not ready for commercial production.

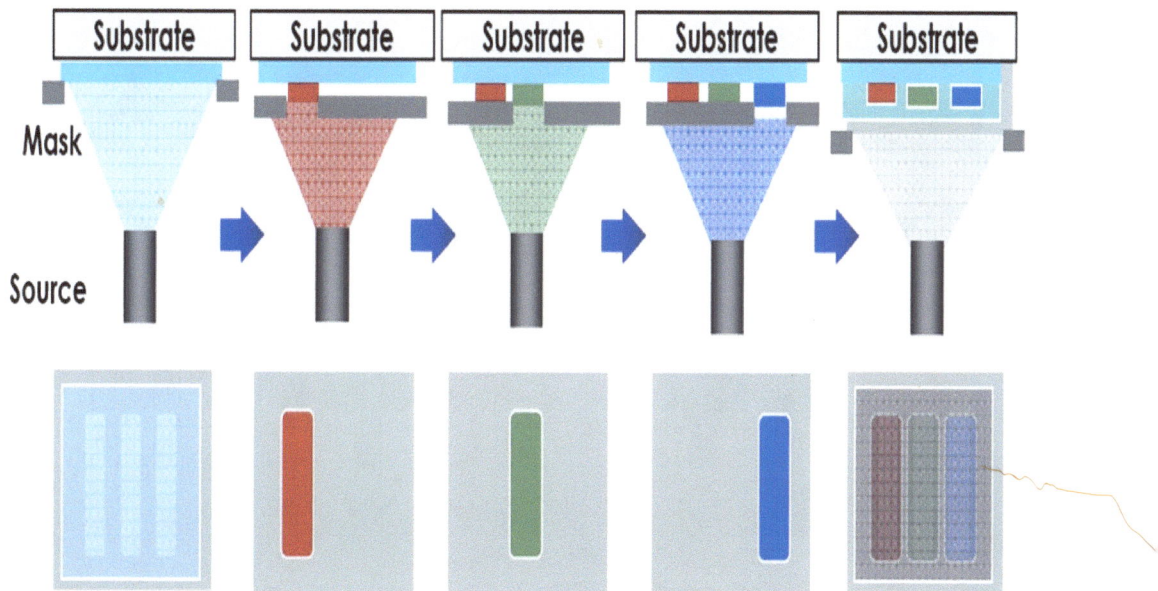

Fig. (2.1). The organic emissive layers are patterned by fine metal masks.

The combination of WOLEDs and color filters was proposed to remedy these deficiencies [5, 10]. As shown in Fig. (**2.2**), in such a scheme, one only needs to pattern the color filters using traditional photolithography technology; thus large area and high resolution can be easily achieved. One of the drawbacks is that most of the white emission (~70%) is absorbed by the color filters, resulting in low panel efficiency. A W-RGBW four primary colors method was proposed to improve the panel efficiency [11 - 14]. It takes the advantage that the W color is used most frequently in practice due to un-saturation of the color of most images, and because the W color is not filtered, the panel efficiency is hence much higher than that of W-RGB method [14]. Assuming the area of each sub-pixel is 1/3, and the transmission of color filters is also 1/3, then the transmitted light in W-RGB scheme is 1/3×1/3(R) + 1/3×1/3(G) + 1/3×1/3(B) =1/3, while for W-RGBW scheme, the transmitted light is ¼×1/3(R) + ¼×1/3(G) + ¼×1/3(B) + ¼×1 (W) =1/2; thus only half of the light is absorbed by the color filters in the W-RGBW scheme. Kodak and LG Display are in favor of this technology. LG Display focuses on making large area OLED-TV by adopting the W-RGBW color patterning scheme; they demonstrated the world's largest 55 inch OLED-TV in Jan 2012 in CES and plan to commercialize it in the second quarter of 2012 in their 8G pilot line [15].

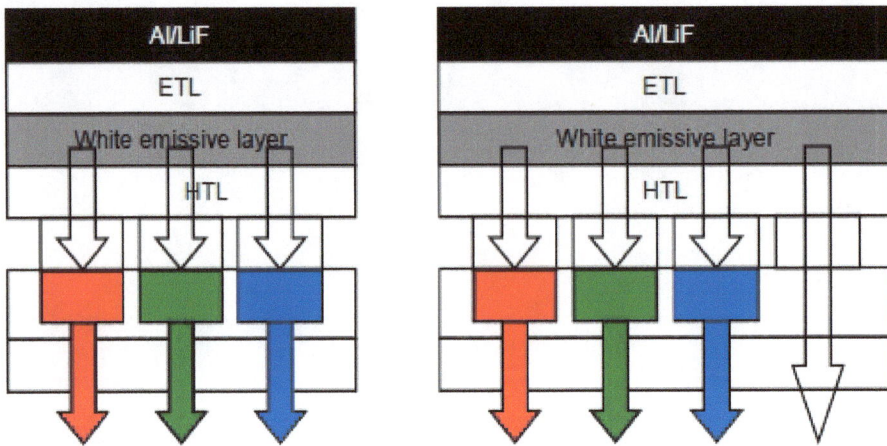

Fig. (2.2). Left: W-RGB, right: W-RGBW color patterning scheme.

A comparison of various color patterning technologies is listed in Table (**2.1**). As can be seen from Table **2.1**, the W-CF patterning method offers highest patterning accuracy/ resolution, largest aperture ratio, fastest TACT and highest yield thus lowest manufacture cost among other color patterning technologies. With the continual improvement of white light efficiency, the shadow-mask-free W-RGBW method is very competitive for low cost, large area and high resolution displays. For example, LG demonstrated the world's largest AMOLED panel by adopting the W-RGB method very recently (shown in Fig. (**2.3**) left) [16]. Again with the W-RGBW technology, e-Magin has commercialized a very high resolution (2128 ppi, 1270×1024×4) microdisplay with pixel size as small as 12 μm in 2008 (shown in Fig. (**2.3**) right) [17].

Fig. (2.3). Left: world's largest 55' AMOLED by LG; right: world's highest resolution AMOLED by eMagin. Both of them are featured with W+CF technology [16, 17].

Table 2.1. Comparisons of various color patterning technologies [5].

	FMM	White/CF	Ink-jet printing	LITI
1.Process aspect				
Patterning accuracy	±20μm	±2.5μm	±15μm	±2.5μm
Resolution	Low, < 200 ppi	High, < 200 ppi	Medium, ~ 200 ppi	High, > 200 ppi
Glass size	Small	Large	Large	Medium
Material usage	~4%	~5%	~80%	-
Material Tact	Worse	Better	Better	Worse
2. Panel properties aspects				
Aperature ratio (top-emission)	40-5-%	~80&	~60&	70-80&
Yield	Low	Excellent	Middle	Low
Panel cost	Expensive	Cheap	Middle	Middle
3. OLED materials aspects				
Material	Small molecule	Small molecule	Ploymer	polymer & SM
Lifetime & efficiency	Better	Better	Worse	Worse
Color Gamut	Good	Good	Bad	Good

WOLEDs FOR SOLID-STATE LIGHTING

The first incandescent light bulb was invented by Thomas Edison in 1880, since then, lighting has become an indispensable commodity in our daily life. Survey shows that about 12% of China's total electricity is consumed by lighting. Most of the lighting sources are dominated by the low efficient incandescent bulbs, which converted approximately 90% of the power as heat rather than as visible light. In order to save energy, China, which has the largest number of people all over the world, is going to prohibit the entrance and sales of the incandescent bulbs starting from Oct. 1, 2012 [18]. By phasing out the inefficient incandescent lamps, 48 billion kilowatt hours of power will be saved annually, and it will also reduce 48 million tons carbon dioxide emission per year. United States and European Union also plan to phase out the bulbs and popularize LED lamps and fluorescent tubes due to their relatively high luminous efficiencies than incandescent bulbs. However, most of the fluorescent lamps, claiming to have an efficiency of 70 lm/W, are actually not that high. When assembled as a luminaire, the actual efficiency is greatly reduced to 30-40 lm/W [19]. In addition, the fluorescent tubes do contain mercury, though little in amount, still cause environmental damage when broken or at their disposal. The inorganic LEDs, although environmentally friendly, suffer from a series of problems such as heat dissipation, high cost per lumen, and cannot serve as true surface-source illumination [19, 20].

The efficiency of WOLEDs has been greatly improved in recent years due to substantial improvement of materials, device structure and light-outcoupling mechanisms [19 - 25]. While the first reported WOLED exhibited a low efficiency of 1 lm/W [1, 2], today's laboratory devices shows an efficiency higher than the 100 lm/W mark [26], rendering WOLEDs very competitive as the new generation lighting source. The greatest advantages of WOLEDs over those of fluorescent tubes or LEDs are that the WOLEDs are a surface light-source, which intrinsically emit the diffusing light, thus eliminating light distribution luminaire as usually employed in fluorescent tubes or LED lamps. Also, since the WOLED is an area source, the generated heat is well distributed in a large volume rather than concentrated on a small point as in the case in LEDs, thus heat dissipation is not as demanding as LEDs. Furthermore, the OLEDs can have the spectra covering a wide range of visible spectrum as shown in Fig. (**2.4**), implying a high CRI of 100 is possible. More importantly, its unique form factors like being ultrathin, light-weight, flexible, transparent, or reflective allow architects and de-signers to introduce light into building in a new and more elegant ways, which shall revolutionize the way light sources exist. Table **2.2** compares the key characteristics of various lighting sources.

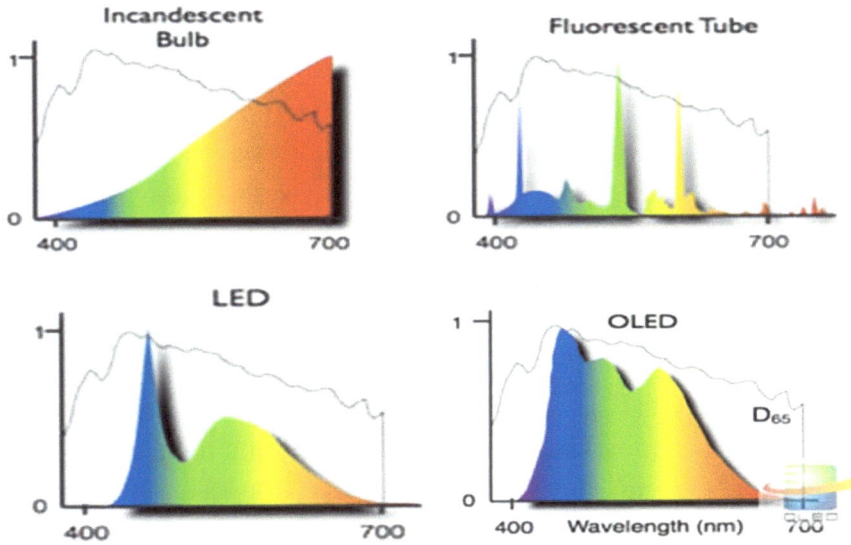

Fig. (2.4). Spectra of various light sources.

Table 2.2. Comparison of Various Light Sources.

Type	lm/W	CCT (K)	CRI	Lifetime (h)
Incandescent bulb	15	2854	100	1500
CFL (warm)	60	2940	82	10000
LED	90	3000	80	60000
WOLED panel [27]	58	2580	83	30000
WOLED pixel [26]	102	3900	70	-

APPROACHES TO WHITE LIGHT EMISSION

In Kido's pioneering work [1], three emitters each emitting blue, green, or red light, were doped into a host matrix to generate white light. The first reported WOLED exhibited a low efficiency of 1 lm/W, but this value has been booted to larger than 100 lm/W over the last few years due to the continual improvement of material performance and device structures. This section reviews several typical device architectures for efficient small molecular WOLEDs.

Multi-emissive Layers

In the multi-emissive structure, a host-guest system is generally adopted to reduce the aggregation-caused quenching effect. By stacking two or three emissive layers with each layer emitting complementary or primary color, a white emission can be realized [28 - 34]. The concentration of excitons in any one of the emitting layers can be managed by varying layer thickness, by adjusting dopant concentration and/or by introducing charge-blocking layers [19]. By tuning the concentration of exctions within individual emitting layers, emission from blue-, green, and red-emitting layers can be balanced and hence white light of the desired color purity can be obtained. For example, Fig. (**2.5**) shows a WOLED structure with three primary color dopants Fir6 (Bis(2,4-difluorophenylpyridinato) tetrakis(1-pyrazolyl) borateiridium(III)), Ir(ppy)$_3$ (Tris(2-phenylpyridine)iridium(III)) and PQIr (Tris(1-phenylisoquinoline) iridium(III)) doped into their host UGH2 (1,4-Bis(triphenylsilyl)benzene), mCP (1,3-Bis(carbazol-9-yl)benzene) and TCTA (4,4',4'-Tri(9-carbazoyl)triphenylamine) as a blue, a green and a red emission layer, respectively. By carefully tuning each emissive layer thickness and dopant concentration, balanced white emission with CIE coordinate of (0.37, 0.41) at 1 mA/cm^2, CRI of 81 and peak efficiency of 38 lm/W have been obtained [28].

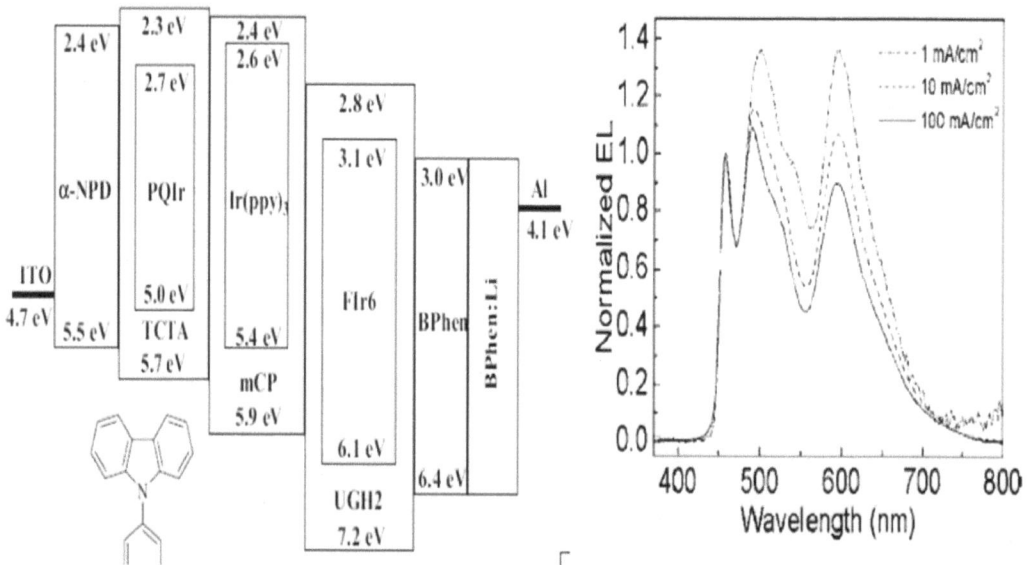

Fig. (2.5). Left: device structure of the multi-emissive WOLEDs; right: spectra of the resulting WOLEDs [28].

Although many efficient WOLEDs have been reported by employing this method, there are several disadvantages. First, the driving voltage is relatively high because the emission region containing many emitting layers is quite thick. The high driving voltage leads to the reduction of the power efficiency. To reduce the driving voltage, thin emissive layers combined with exciton-confinement layer, or doped carrier transport layer are proposed [34].

In addition, the emission color changes at different driving voltage/ current as shown in Fig. (**2.5**), which is ascribed to the shifting of the recombination zone. Furthermore, differential aging of the color dopants also leads to undesired color shift as the device ages. Finally, the fabrication process is relatively complex and the devices involve lots of functional layers which may increase the cost.

Single-emissive Layer

In the single-emissive layer structure, several color dopants are doped simultaneously into a single host material [35 - 40]. By carefully tuning each dopant concentration, a balanced white emission can be achieved. For instance, D'Andrade *et al.* demonstrated an efficient WOLED by employing three color dopants Fir6, $Ir(ppy)_3$ and PQIr doped into a wide energy band-gap host UGH2 with a doping concentration of 20 wt.%, 0.5 wt.% and 2 wt.%, respectively [35]. As shown in Fig. (**2.6**), the TCTA and TPBi (2,2',2"-(1,3,5-Benzinetriyl)-tri-(1-phenyl-1-H-benzimidazole)) were employed as exciton confinement layers to confine the recom-bination within the UGH2 layer. Due to the relatively wide energy band-gap of UGH2 than that of Fir6, as well as the highest doping concentration of Fir6 among the dopants, the injected holes and electrons are directly captured by Fir6 and subsequently form the excitons on Fir6. Exciton direct formation avoids the losses of exchange energy (0.5-1 eV), which otherwise commonly occurs in the host-guest system during exciton energy transfer from the wide energy host to the low energy guest. This leads to a reduction of driving voltage and hence an increase of power efficiency. By subsequently transferring a part of the Fir6 exciton energy to the red and green dopants *via* Forster or Dexter energy transfer, a red and a green emission can be obtained. The mixing of the R, G, and B emission results in a balanced white emission with efficiency of 14 lm/W at 10 mA/cm^2, CRI of 80 and CIE coordinates of (0.38, 0.45).

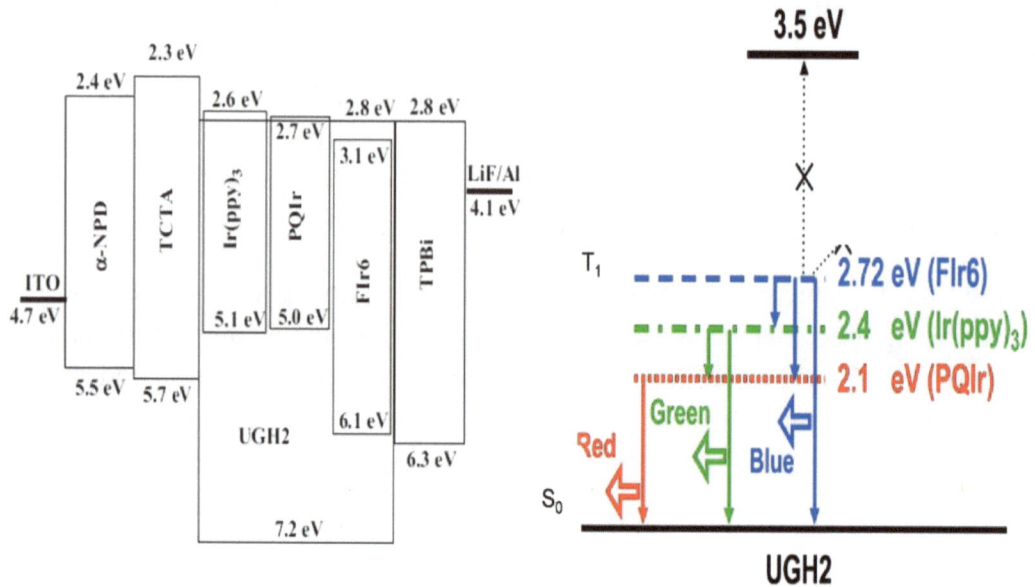

Fig. (2.6). Device structure for WOLEDs with a tri-doped single emissive layer (left) and energy transfer in such a single emissive layer (right) [35].

In such triple-doped WOLDs structure, differential aging of each dopant can be eliminated as only the blue dopant conducts the charges, and is the sole site for direct exciton formation. As blue emission decreases due to degradation, the green and red emission should decrease correspon-dingly because their emission intensities are directly related to that of the blue emission. The single emissive layer structure is commonly adopted for fabricating polymer WOLEDs using solution process due to the difficulty of preparing multilayer in the solution process [39, 40]. For small molecular OLEDs fabricated by thermal evaporation, the triple- or duple-doped process may be somewhat complicated, as one has to simultaneously evaporate three to four kinds of materials and maintain the dopants deposition rate steadily.

WOLEDs with Fluorescent-phosphorescent Hybrid Emitters

To make a high efficiency WOLEDs, phosphorescent emitters that emit red, green and blue colors are adopted due to their theoretically high internal quantum efficiency of 100%. However, it is difficult to find saturated blue phosphorescent emitters which are sufficiently stable. A hybrid structure combining a fluorescent blue emitter with a phosphorescent red and a phosphorescent green emitter was proposed for simultaneously achieving high efficiency and long operational lifetime [41 - 44]. It was shown that theoretically a 100% internal quantum efficiency can be reached in such a hybrid structure if triplet and singlet excitons are distributed in an appropriate way. Such a hybrid structure takes the advantage that the warm white light contains 25% blue emission and the devices also contain 25% singlet excitons. If the 25% singlet excitons recombine in the fluorescent blue emitter, and the 75% triplet excitons recombine in the red and the green phosphors, then all excitons can be harvested theoretically.

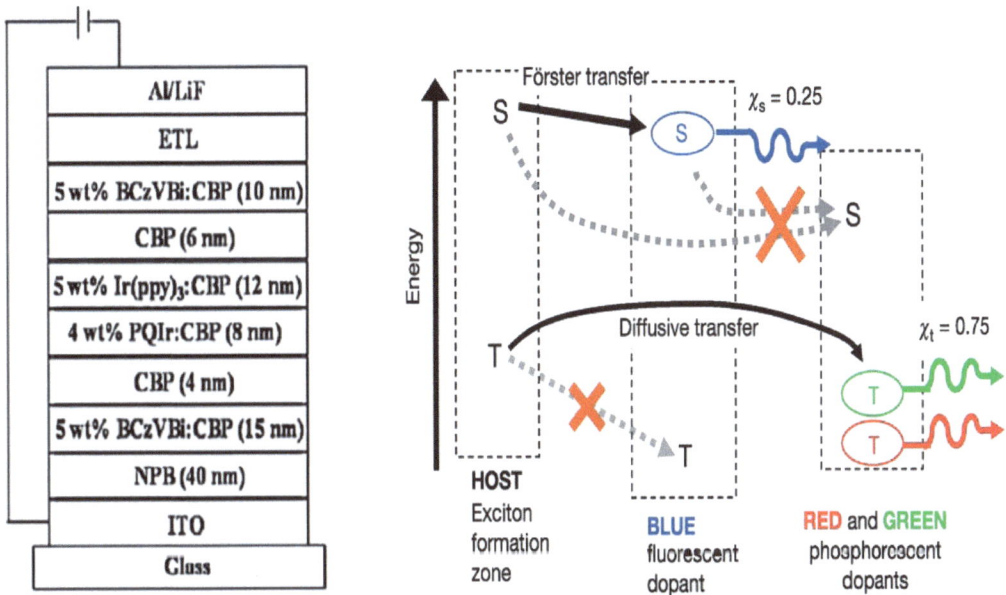

Fig. (2.7). Device structure of a hybrid WOLEDs (left) and energy transfer in such hybrid devices (right) [41].

As shown in Fig. (**2.7**), the singlet exciton energy transfers from the hosts to the fluorescent blue emitter *via* Forster mechanism, while the triplet excitons diffused from the exciton formation zone to the green and the red emission layers and finally transfer their energy to the phosphorescent green and red emitters *via* Dexter mechanism. In order to harvest all excitons, special care must be taken to ensure all singlet excitons are confined and recombined within the fluorescent blue emitter. To meet such requirement, the fluorophor should be placed within the exciton generation zone, whereas the phosphorescent emitters are placed at a greater distance far away from the exaction generation zone. An interlayer with thickness larger than the singlet diffusing length (<5 nm) but smaller than the triplet diffusing length is often inserted between the fluorophor and phosphors, so that the singlet cannot reach the phosphors whereas the triplet can still diffuse to the phosphors due to their long lifetime.

Also, the triplet energy level of the fluorophor should be higher than that of the hosts and phosphors to prevent triplet excitons trapped by the fluorophor. In the pioneering work by Sun *et al.* the hybrid WOLEDs exhibited an efficiency of 14 lm/W at a luminance of 500 cd/m^2 with a CRI of 85 and a warm white CIE coordinates of (0.40, 0.41) [41]. By employing a high triplet energy level blue emitter, K. Leo's group demonstrated an improved efficiency of 22 lm/W at 1000 cd/m^2 [42].

Tandem WOLEDs

Instead of stacking several color emissive layers or doping the color dopants into one single host, one can also connect several OLEDs in series with each OLED emitting one primary or complementary colors to generate the white emission [45 - 50]. A typical tandem structure is shown in Fig. (**2.8**). In this tandem structure, an injected electron can be used to generate multi-photons, resulting in a very high luminance and thus high current efficiency. However, the driving voltage is also increased, and thus the power efficiency of the tandem devices is not higher than that of single-cell devices. To construct a tandem WOLED, a transparent internal electrode often referred as charge generation layer (CGL) is employed. The CGL generally consists of a n-doped electron injection layer such as Li:Bphen (4,7-Diphenyl-1,10-phenanthroline) and a hole injection layer like MoO$_3$, WO$_3$, or p-doped NPB (N,N' -Bis(naphthalen-1-yl)-N,N' -bis(phenyl- -benzidine). Such composite layers can therefore provide electrons to one subcell and holes to another subcell. For example, with a MoO3/Li:Bphen CGL, Forrest's group have demonstrated a stacked WOLED with efficiency of 13 lm/W at 1000 cd/m^2, CIE coordinates of (0.45, 0.36) and CRI of 63 [45].

One of the advantages of such method is that the chromaticity and CRI of the emission color can be tuned over a wide range, given that each subcell can be addressed independently [51]. Furthermore, the color shift due to differential aging of each subcell can be compensated by tuning the driving current of each subcell. More importantly, the lifetime of tandem devices is significantly higher than that of signal-cell devices due to the reduction of driving current. The disadvantage of this method is the relatively complex process due to the large number of layers involved, hence potentially costly.

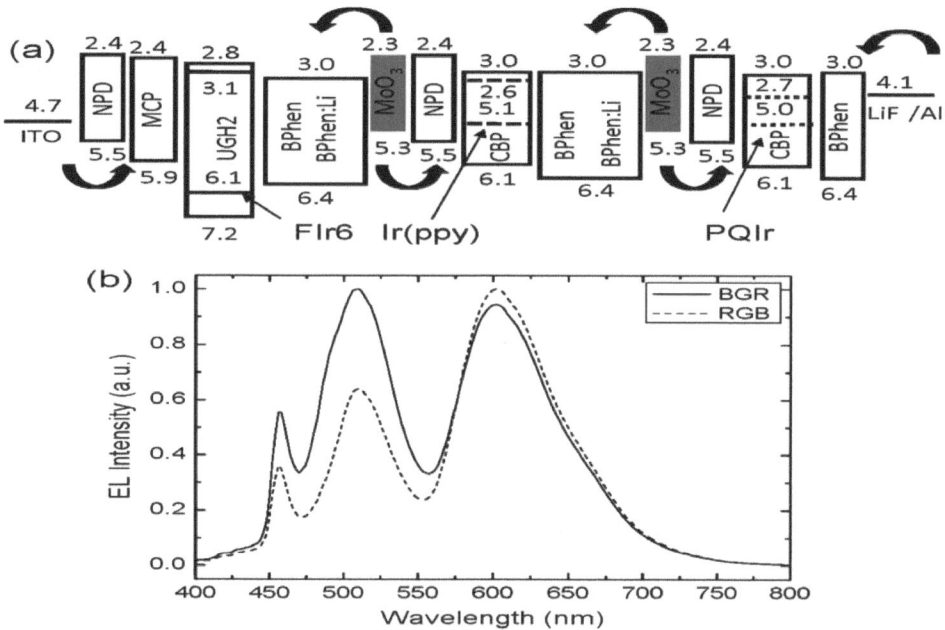

Fig. (2.8). Device structure of a tandem WOLED (**a**) and spectra of the resulting WOLED (**b**) [45].

Side by Side WOLEDs

Instead of arranging several color OLEDs vertically in the tandem structure, one may also arrange the color OLEDs horizontally, as shown in Fig. (**2.9**). If the distance between adjacent OLEDs is close enough, then the mixing of the emission from the color OLEDs can generate a white emission. Same as the stacked devices with separated contacts/ electrodes for each subcell, color adjustment by independently changing the intensity of blue, green and red emission can be easily achieved.

The performance of the blue, green and red OLEDs can also be optimized independently. The main issue of this method is the relative elaborate fabrication process as the organic emissive layers should be patterned by expensive fine shadow masks.

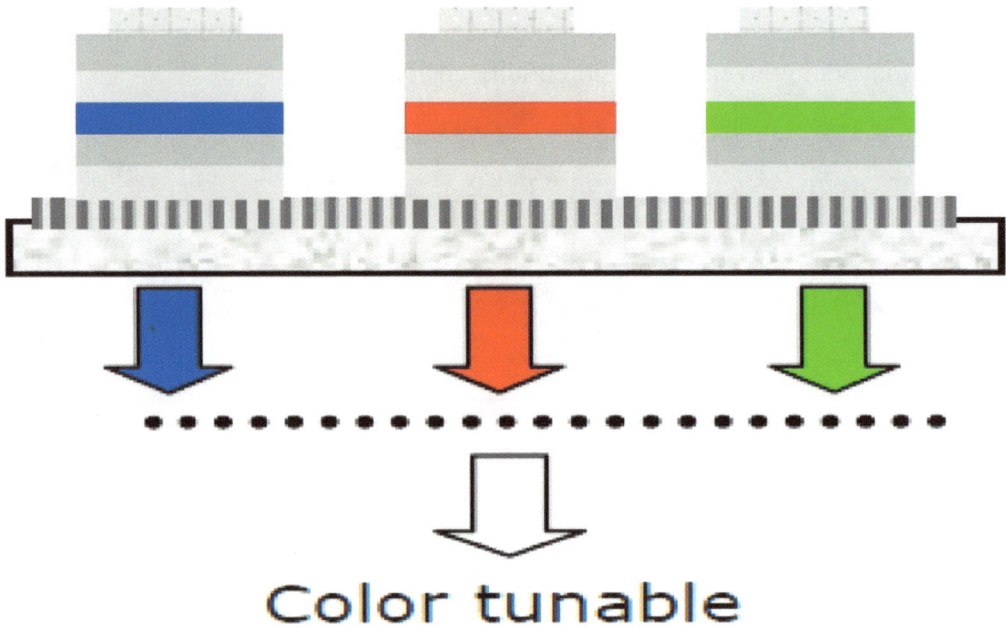

Fig. (2.9). A color tunable striped type OLED.

Color Converted WOLEDs

As shown in Fig. (**2.10**), the WOLEDs can also be constructed by depositing green/red phosphors on the back substrate side of the blue OLEDs [52 - 55]. These phosphors serve as external color conversion media, which absorb a part of the blue photons and subsequently convert them to their own green/red emission *via* a photoluminescence process. The mixing of the converted green/red emission and the unabsorbed blue emission generates the white emission. Using this strategy, WOLEDs with high CRI of 93 and efficiency of 3.8 lm/W have been reported [52]. The biggest advantage of such WOLEDs is that the color shift either due to differential aging of emitters or shifting of the recombination zone, as is usually the case in conventional WOLEDs, is completely eliminated. This is because the intensity of the converted red/green emission is always proportional to the blue emission and thus ultra high color stability is achieved. Also, the fabrication process is much simpler compared with that of other types of WOLEDs, which generally involve several doping layers and require careful tuning of layer thickness and dopant concentration.

One of the drawbacks is that the efficiency of the color converted WOLEDs strongly depends on that of the blue OLEDs, while the blue emitters show lowest efficiency and reliability currently. Also, energy loss is inevitable during the blue-to-green/red color conversion process.

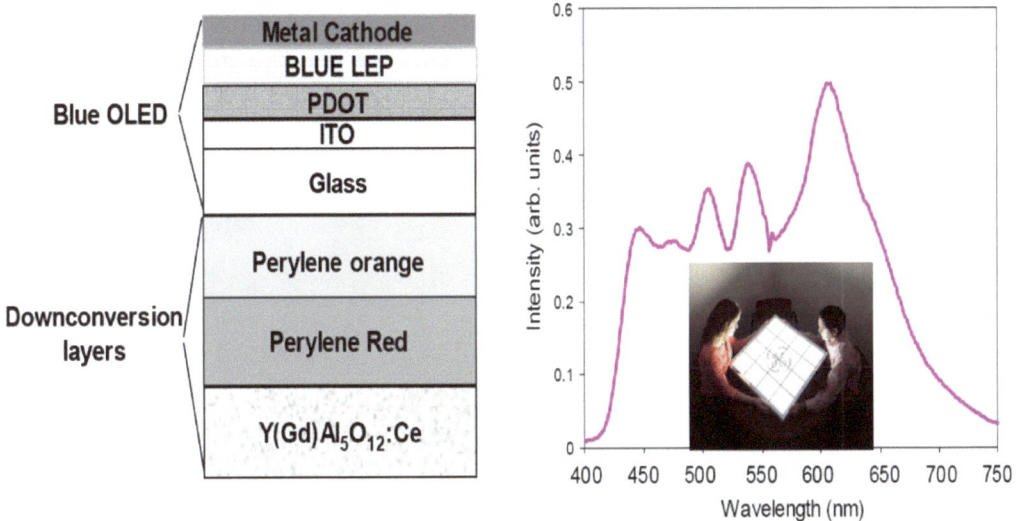

Fig. (2.10). Device structure (left) and spectrum (right) of the color converted WOLEDs [53].

Excimer/Exciplex WOLEDs

An excimer/exciplex as shown in Fig. (**2.11**) is a dimer formed because of bimolecular interaction between the ground state of one molecule with the excited states of another molecule. Many square-planar Pt complexes tend to form excimer at high doping concentration [56 - 60]. When the excimer returns to the ground state, its components dissociate and often repel each other, accompanied with a considerably red shifted and broadened emission as compared with the monomer emission.

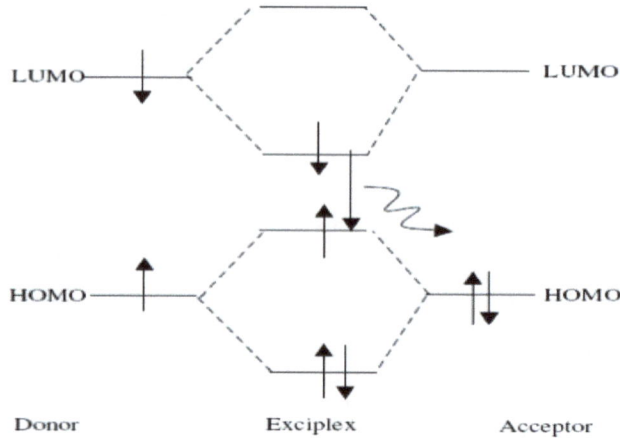

Fig. (2.11). Schematic illustration of the emission due to the formation of exciplexes/excimers.

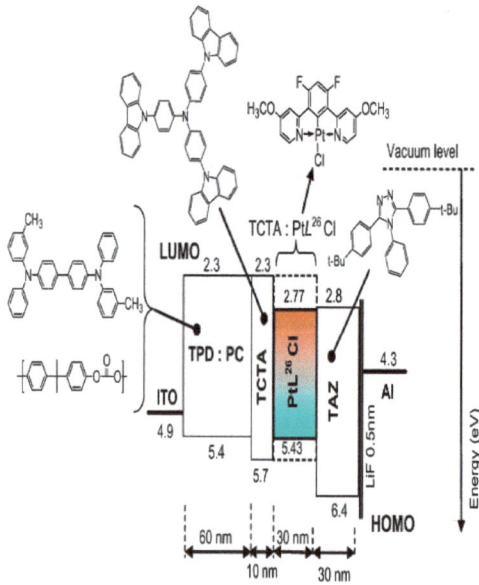

Fig. (2.12). Device structure (left) and spectra (right) of the excimer OLEDs [56].

By mixing the red shifted excimer emission with the monomer emission, a balanced white emission can be obtained [56 - 60]. For example, Cocchi *et al.* demonstrated that a wide range of color from blue to red can be obtained by simply tuning the concentration of a Pt complex in a simple tri-layer device structure. As shown in Fig. (**2.12**), at low doping concentration of 5%, a blue emission originating from monomer emission with efficiency of 2.8 lm/W was obtained. Further increasing the doping concentration to 100%, a red color due to excimer emission was achieved with efficiency of 8.1 lm/W.

By tuning the relative intensity of the red excimer emission and the blue monomer emission, very pure white emission with CRI of 74, CIE coordinates of (0.34, 0.45), and efficiency of 6.8 lm/W were obtained at a doping concentration of 20% [56]. The excimer offers a very simple approach for realizing color stable white emission in a simple device structure; for example, the emission can cover the entire visible spectrum by using only one or at most two dopants.

Table 2.3. Comparisons of various WOLED structures.

Device structure	Efficiency [a]	CRI	CIE	No. of layers/ materials	Fabrication process	Color shift due to differential aging	Color shift due to differential brightness	Other comments	Ref.
Multi-emissive	55.2 lm/w at 100 cd/m^2	82	(0.40, 4.40)	6/8	Complex	Y	Y	High efficiency	[29]
Single-emissive	19 lm/W at 500 cd/m^2	-	(0.33, 0.39)	4/7	Complex	Y	Y	Good reliability	[38]
Hybrid	22 lm/W at 500 cd/m^2	85	(0.40, 0.41)	7/6	Complex	Y	Y	Good Realibility	[42]
Tandem	13 lm/W at 1000 cd/m^2	63	0.35, 0.46)	14/10	Quite Complex	Y	little	Good reliability	[45]
side-by side	-	-	-	-	Very complex	Can be compensated	N	Color tunale/high efficiency	-
Color converted	15 lm/W at 1000 cd/m^2	88	(0.36, 0.36)	5/6	Simple	N	N	Efficiency/stability depend on blue emitter	[53]
Excimer	12.6 lm/W at 500 cd/m^2	69	0.46, 0.47)	4/5	Simple	N	N	Structure simplicity	[59]

[a] all efficiencies listed here do not consider enhancement by using light outcoupling techniques. If light outcoupling methods are employed, an enhancement factor of 2-3 should be multiplied.

The advantages and disadvantages of various WOLEDs architectures are compared in Table (**2.3**). For clear comparison, all efficiencies extracted from their corresponding literatures do not consider the enhancement by employing light out coupling techniques. These efficiencies represent the typical values for the state-of-the-art WOLEDs. If certain light out-coupling techniques like high refractive index substrate and microlens array are employed, an enhancement factor of 2-3 can be achieved, indicating white efficiency larger than 100 lm/W is possible.

CONFLICT OF INTEREST

The authors confirm that this chapter contents have no conflict of interest.

ACKNOWLEDGMENTS

This work was supported by the National Natural Science Foundation of China under Grant No. 61405089, and by the Innovation of Science and Technology Committee of Shenzhen under Grant No. JCYJ20140417105742713.

REFERENCES

[1]   J. Kido, K. Hongawa, K. Okuyama, and K. Nagei, "White lightemitting organic electroluminescent devices using the poly(Nvinylcarbazole) emitter layer doped with three fluorescent dyes", *Appl. Phys. Lett.,* vol. 64, p. 815, 1994.
        [http://dx.doi.org/10.1063/1.111023]

[2]   J. Kido, M. Kimura, and K. Nagai, "Multilayer white light-emitting organic electroluminescent device", *Science,* vol. 267, no. 5202, pp. 1332-1334, 1995.
        [http://dx.doi.org/10.1126/science.267.5202.1332] [PMID: 17812607]

[3]   J-J. Lih, C-I. Chao, and C-C. Lee, "Novel pixel design for high-resolution AMOLED displays with a shadow mask", *J. Soc. Inf. Disp.,* vol. 15, pp. 3-7, 2007.
        [http://dx.doi.org/10.1889/1.2451539]

[4]   H.D. Kim, J.K. Jeong, H.-J. C, and Y.-G. Mo, "22.1: Invited Paper: Technological Challenges for Large-Size AMOLED Display", *SID Symp. Dig. Tech. Papers,* vol. 39, pp. 291-294, 2008.

[5]   K. Chung, N. Kim, J. Choi, C. Chu, and J-M. Huh, "70.1: Invited paper: large-sized full color AMOLED TV:advancements and issues", *SID Symp. Dig. Tech. Papers,* vol. 37, pp. 1958-1963, 2006.
        [http://dx.doi.org/10.1889/1.2451418]

[6]   T. Gohda, Y. Kobayashi, K. Okano, S. Inoue, K. Okamoto, S. Hashimoto, E. Yamamoto, H. Morita, S. Mitsui, and M. Koden, "58.3: A 3.6-in. 202-ppi full-Color AMPLED display fabricated by ink-jet method", *SID Symp. Dig. Tech. Papers,* vol. 37, pp. 1767-1770, 2006.
        [http://dx.doi.org/10.1889/1.2433379]

[7]   N.C. van der Vaart, H. Lifka, F.P. Budzelaar, J.E. Rubingh, J.J. Hoppenbrouwers, J.F. Dijksman, R.G. Verbeek, R. van Woudenberg, F.J. Vossen, M.G. Hiddink, J.J. Rosink, T.N. Bernards, A. Giraldo,

N.D. Young, D.A. Fish, M.J. Childs, W.A. Steer, D. Lee, and D.S. George, "Towards large-area full-color active-matrix printed polymer OLED television", *J. Soc. Inf. Disp.*, vol. 13, pp. 9-16, 2005.
[http://dx.doi.org/10.1889/1.1867094]

[8]   T. Hirano, K. Matsuo, K. Kohinata, K. Hanawa, T. Matsumi, E. Matsuda, R. Matsuura, T. Ishibashi, A. Yoshida, and T. Sasaoka, "53.2: distinguished paper: novel laser transfer technology for manufacturing large-sized OLED displays", *SID Symp. Dig. Tech. Papers,* vol. 38, pp. 1592-1595, 2007.
[http://dx.doi.org/10.1889/1.2785623]

[9]   M. Boroson, L. Tutt, K. Nguyen, D. Preuss, M. Culver, and G. Phelan, "16.5L: Late-News-Paper: non-contact OLED color patterning by radiation-induced sublimation transfer (RIST)", *SID Symp. Dig. Tech. Papers,* vol. 36, pp. 972-975, 2005.
[http://dx.doi.org/10.1889/1.2036612]

[10]   S. Kim, S. Lee, M. Kim, J. Song, E. Hwang, S. Tamura, S. Kang, H. Kim, C. Kim, J. Lee, J. Kim, S. Cho, J. Cho, M. C. Suh, and H. Kim, "A 3.0-in. 308-ppi WVGA AMOLED with a top-emission white OLED and color filter", *J. Soc. Inf. Disp.,* vol. 17, pp. 145-149, 2009.
[http://dx.doi.org/10.1889/JSID17.2.145]

[11]   A. D. Arnold, P. E. Castro, T. K. Hatwar, M. V. Hettel, P. J. Kane, and J. E. Ludwicki, "Full-color AMOLED with RGBW pixel pattern", *J. Soc. Inf. Disp.,* vol. 13, pp. 525-535, 2005.
[http://dx.doi.org/10.1889/1.1974009]

[12]   J.P. Spindler, T.K. Hatwar, M.E. Miller, A.D. Arnold, and S.A. VanSlyke, "System considerations for RGBW OLED displays", *J. Soc. Inf. Disp.,* vol. 14, pp. 37-48, 2006.
[http://dx.doi.org/10.1889/1.2166833]

[13]   B-W. Lee, K. Park, A. Arkhipov, S. Shin, and K. Chung, "L-2: Late-News Paper: the RGBW advantage for AMOLED", *SID Symp. Dig. Tech. Papers,* vol. 38, pp. 1386-1389, 2007.
[http://dx.doi.org/10.1889/1.2785572]

[14]   J.W. Hamer, A.D. Arnold, M. Itoh, T.K. Hatwar, M.J. Helber, and S.A. VanSlyke, "System design for a wide-color-gamut TV-sized AMOLED display", *J. Soc. Inf. Disp.,* vol. 16, pp. 3-14, 2008.
[http://dx.doi.org/10.1889/1.2835033]

[15]   http://www.osa-direct.com/ osad-news/lg-display-to- launch-large-area-oled-tv-in-2012.html.

[16]   http://ces.cnet.com/ 8301-33379_1-57358177/ lgs-55-inch-55em9600-oled-tv-wins-best-of-ces/.

[17]   http://www.emagin.com/oled-microdisplays/.

[18]   http://news.xinhuanet.com/fortune/ 2011-11/04/c_111146231.htm.

[19]   K.T. Kamtekar, A.P. Monkman, and M.R. Bryce, "Recent advances in white organic light-emitting materials and devices (WOLEDs)", *Adv. Mater.,* vol. 22, no. 5, pp. 572-582, 2010.
[http://dx.doi.org/10.1002/adma.200902148] [PMID: 20217752]

[20]   B.W. D'Andrade, and S.R. Forrect, "White organic light-emitting devices for solid-state lighting", *Adv. Mater.,* vol. 16, p. 1585, 2004.
[http://dx.doi.org/10.1002/adma.200400684]

[21]   L. Xiao, Z. Chen, B. Qu, J. Luo, S. Kong, Q. Gong, and J. Kido, "Recent progresses on materials for electrophosphorescent organic light-emitting devices", *Adv. Mater.,* vol. 23, no. 8, pp. 926-952, 2011.

[http://dx.doi.org/10.1002/adma.201003128] [PMID: 21031450]

[22] G. Zhou, W-Y. Wong, and S. Suo, "Recent progress and current challenges in phosphorescent white organic light-emitting diodes (WOLEDs)", *J. Photochem. Photobiol. Photochem. Rev.,* vol. 11, pp. 133-156, 2011.
[http://dx.doi.org/10.1016/j.jphotochemrev.2011.01.001]

[23] M.C. Gather, A. Köhnen, and K. Meerholz, "White organic light-emitting diodes", *Adv. Mater.,* vol. 23, no. 2, pp. 233-248, 2011.
[http://dx.doi.org/10.1002/adma.201002636] [PMID: 20976676]

[24] D. Gupta, and M. Katiyar, "Various approaches to white organic light emitting diodes and their recent advancements", *Opt. Mater.,* vol. 28, pp. 295-301, 2006.
[http://dx.doi.org/10.1016/j.optmat.2004.11.036]

[25] T. Tsuboi, "Recent advances in white organic light emitting diodes with a single emissive dopant", *J. Non-Cryst. Solids,* vol. 356, pp. 1919-1927, 2010.
[http://dx.doi.org/10.1016/j.jnoncrysol.2010.05.034]

[26] http://www.universaldisplay.com/downloads/Press%20Releases/2008/PANL_white milestone_FINAL.pdf.

[27] http://www.oled-info.com/udc-unveils-new-oled- lighting-panel-58-lmw-30000-lifetime-hours.

[28] Y. Sun, and S.R. Forrest, "High-efficiency white organic light emitting devices with three separate phosphorescent emission layers", *Appl. Phys. Lett.,* vol. 91, p. 263503, 2007.
[http://dx.doi.org/10.1063/1.2827178]

[29] H. Sasabe, J. Takamatsu, T. Motoyama, S. Watanabe, G. Wagenblast, N. Langer, O. Molt, E. Fuchs, C. Lennartz, and J. Kido, "High-efficiency blue and white organic light-emitting devices incorporating a blue iridium carbene complex", *Adv. Mater.,* vol. 22, no. 44, pp. 5003-5007, 2010.
[http://dx.doi.org/10.1002/adma.201002254] [PMID: 20824668]

[30] G. Cheng, Y. Zhang, Y. Zhao, Y. Lin, C. Ruan, S. Liu, T. Fei, Y. Ma, and Y. Cheng, "White organic light-emitting devices with a phosphorescent multiple emissive layer", *Appl. Phys. Lett.,* vol. 89, p. 043504, 2006.
[http://dx.doi.org/10.1063/1.2227645]

[31] M-H. Ho, S-F. Hsu, and J-W. Ma, "'White p-i-n organic light-emitting devices with high power efficiency and stable color", *Appl. Phys. Lett.,* vol. 91, p. 113518, 2007.
[http://dx.doi.org/10.1063/1.2784971]

[32] M-F. Lin, L. Wang, W-K. Wong, K-W. Cheah, H-L. Tam, M-T. Lee, M-H. Ho, and C.H. Chen, "Highly efficient and stable white light organic light-emitting devices", *Appl. Phys. Lett.,* vol. 91, p. 073517, 2007.
[http://dx.doi.org/10.1063/1.2769762]

[33] C-H. Chang, C-C. Chen, C-C. Wu, S-Y. Chang, J-Y. Hung, and Y. Chi, "High-color-rendering pure-white phosphorescent organic light-emitting devices employing only two complementary colors", *Org. Electron.,* vol. 11, pp. 266-272, 2010.
[http://dx.doi.org/10.1016/j.orgel.2009.11.004]

[34] S. Reineke, F. Lindner, G. Schwartz, N. Seidler, K. Walzer, B. Lüssem, and K. Leo, "White organic

light-emitting diodes with fluorescent tube efficiency", *Nature,* vol. 459, no. 7244, pp. 234-238, 2009. [http://dx.doi.org/10.1038/nature08003] [PMID: 19444212]

[35]   B.W. D'Andrade, R.J. Holmes, and S.R. Forrest, "Efficient organic electrophosphorscent white-light-emitting device with a triple doped emissive layer", *Adv. Mater.,* vol. 16, p. 624, 2004. [http://dx.doi.org/10.1002/adma.200306670]

[36]   S-H. Eom, Y. Zheng, E. Wrzesniewwski, J. Lee, N. Chopra, F. So, and J. Xie, "White phosphorescent organic light-emitting devices with dual triple-doped emissive layers", *Appl. Phys. Lett.,* vol. 94, p. 153303, 2009. [http://dx.doi.org/10.1063/1.3120276]

[37]   Q. Wang, J. Ding, D. Ma, Y. Cheng, and L. Wang, "Highly efficient single-emitting-layer white organic light-emitting diodes with reduced efficiency roll-off", *Appl. Phys. Lett.,* vol. 94, p. 103503, 2009. [http://dx.doi.org/10.1063/1.3097028]

[38]   Q. Wang, J. Ding, D. Ma, Y. Cheng, L. Wang, X. Jing, and F. Waong, "Harvesting excitons via two parallel channels for efficient white organic LEDs with nearly 100% internal quantum efficiency: fabrication and emission-mechanism analysis", *Adv. Funct. Mater.,* vol. 19, pp. 84-95, 2009. [http://dx.doi.org/10.1002/adfm.200800918]

[39]   Y. Xu, J. Peng, J. Jiang, W. Xu, W. Yang, and Y. Cao, "Efficient white-light-emitting diodes based on polymer codoped with two phosphorescent dyes", *Appl. Phys. Lett.,* vol. 87, p. 193502, 2005. [http://dx.doi.org/10.1063/1.2119407]

[40]   M.C. Gather, R. Alle, H. Becker, and K. Meerholz, "On the origin of the color shift in white-emitting OLEDs", *Adv. Mater.,* vol. 19, pp. 4460-4465, 2007. [http://dx.doi.org/10.1002/adma.200701673]

[41]   Y. Sun, N.C. Giebink, H. Kanno, B. Ma, M.E. Thompson, and S.R. Forrest, "Management of singlet and triplet excitons for efficient white organic light-emitting devices", *Nature,* vol. 440, no. 7086, pp. 908-912, 2006. [http://dx.doi.org/10.1038/nature04645] [PMID: 16612378]

[42]   G. Schwartz, S. Reineke, T.C. Rosenow, K. Walzer, and K. Leo, "Triplet harvesting in hybrid white organic light-emitting diodes", *Adv. Funct. Mater.,* vol. 19, pp. 1319-1333, 2009. [http://dx.doi.org/10.1002/adfm.200801503]

[43]   M.E. Kondakova, J.C. Deaton, T.D. Pawlik, D.J. Giesen, D.Y. Kondakov, R.H. Young, T.L. Royster, D.L. Comfort, and J.D. Shore, "Highly efficient fluorescent-phosphorescent triplet-harvesting hybrid organic light-emitting diodes", *J. Appl. Phys.,* vol. 107, p. 014515, 2010. [http://dx.doi.org/10.1063/1.3275053]

[44]   G. Schwartz, S. Reineke, K. Walzer, and K. Leo, "Reduced efficiency roll-off in high-efficiency hybrid white organic light-emitting diodes", *Appl. Phys. Lett.,* vol. 92, p. 053311, 2008. [http://dx.doi.org/10.1063/1.2836772]

[45]   X. Qi, M. Slootsky, and S.R. Forrest, "Stacked white organic light emitting devices consisting of separate red, green, and blue elements", *Appl. Phys. Lett.,* vol. 93, p. 193306, 2008. [http://dx.doi.org/10.1063/1.3021014]

[46]   T-W. Lee, T. Noh, B-K. Choi, M-S. Kim, D.W. Shin, and J. Kido, "High-efficiency stacked white organic light-emitting diodes", *Appl. Phys. Lett.,* vol. 93, p. 043301, 2008.
[http://dx.doi.org/10.1063/1.2837419]

[47]   Q. Wang, J. Ding, Z. Zhang, D. Ma, Y. Cheng, L. Wang, and F. Wang, "A high-performance tandem white organic light-emitting diode combining highly effective white-units and their interconnection layer", *Appl. Phys. Lett.,* vol. 105, p. 076101, 2009.

[48]   J.X. Sun, X.L. Zhu, H.J. Peng, M. Wong, and H.S. Kwok, "Bright and efficient white stacked organic light-emitting diodes", *Org. Electron.,* vol. 8, pp. 305-310, 2007.
[http://dx.doi.org/10.1016/j.orgel.2006.11.006]

[49]   C-C. Chang, J-F. Chen, S-W. Hwang, and C.H. Chen, "Highly efficient white organic electroluminescent devices based on tandem architecture", *Appl. Phys. Lett.,* vol. 87, p. 253501, 2005.
[http://dx.doi.org/10.1063/1.2147730]

[50]   M-H. Ho, T-M. Chen, P-C. Yeh, S-W. Hwang, and C.H. Chen, "Highly efficient p-i-n white organic light emitting devices with tandem structure", *Appl. Phys. Lett.,* vol. 91, p. 233507, 2007.
[http://dx.doi.org/10.1063/1.2822398]

[51]   Z.L. Shen, P.E. Burrows, V. Bulovic, S.R. Forrest, and M.E. Thompson, "Three-color, tunable, organic light-emitting devices", *Science,* vol. 276, p. 2009, 1997.
[http://dx.doi.org/10.1126/science.276.5321.2009]

[52]   A.R. Duggal, J.J. Shiang, C.M. Heller, and D.F. Foust, "Organic light-emitting devices for illumination quality white light", *Appl. Phys. Lett.,* vol. 80, p. 3470, 2002.
[http://dx.doi.org/10.1063/1.1478786]

[53]   A.R. Duggal, J.J. Shiang, C.M. Heller, D.F. Foust, L.G. Turner, W.F. Nealon, and J.C. Bortscheller, "4.1: invited paper: large area white OLEDs", *SID Symp. Dig. Tech. Papers,* vol. 36, pp. 28-31, 2005.
[http://dx.doi.org/10.1889/1.2036427]

[54]   B.C. Krummacher, V-E. Choong, M.K. Mathai, S.A. Choulis, and F. So, "Highly efficient white organic light-emitting diode", *Appl. Phys. Lett.,* vol. 88, p. 113506, 2006.
[http://dx.doi.org/10.1063/1.2186080]

[55]   Y.B. Yuan, S. Li, Z. Wang, H.T. Xu, and X. Zhou, "White organic light-emitting diodes combining vacuum deposited blue electrophosphorescent devices with red surface color conversion layers", *Opt. Express,* vol. 17, no. 3, pp. 1577-1582, 2009.
[http://dx.doi.org/10.1364/OE.17.001577] [PMID: 19188987]

[56]   M. Cocchi, J. Kalinowski, L. Murphy, J.A. Williams, and V. Fattori, "Mixing of molecular exciton and excimer phosphorescence to tune color and efficiency of organic LEDs", *Org. Electron.,* vol. 11, pp. 388-396, 2010.
[http://dx.doi.org/10.1016/j.orgel.2009.11.017]

[57]   M. Cocchi, J. Kalinowski, V. Fattori, J.A. Williams, and L. Murphy, "Color-variable highly efficient organic electrophosphorescent diodes manipulating molecular exciton and excimer emissions", *Appl. Phys. Lett.,* vol. 94, p. 073309, 2009.
[http://dx.doi.org/10.1063/1.3086900]

[58]   X. Yang, Z. Wang, S. Madakuni, J. Li, and G.E. Jabbour, "Highly efficient excimer-based white

phosphorescent devices with improved power efficiency and color rendering index", *Appl. Phys. Lett.,* vol. 93, p. 193305, 2008.
[http://dx.doi.org/10.1063/1.3013324]

[59] E.L. Williams, K. Haavisto, J. Li, and G.E. Jabbour, "Excimer-based white phosphorescent organic light emitting diodes with nearly 100%internal quantum efficiency", *Adv. Mater.,* vol. 19, pp. 197-202, 2007.
[http://dx.doi.org/10.1002/adma.200602174]

[60] X. Yang, F-I. Wu, H. Haverinen, J. Li, C-H. Cheng, and G.E. Jabbour, "Efficient organic light-emitting devices with platinum-complex emissive layer", *Appl. Phys. Lett.,* vol. 98, p. 033302, 2011.
[http://dx.doi.org/10.1063/1.3541447]

CHAPTER 3

Light Outcoupling Technologies

Yibin Jiang[*], Shuming Chen

Department of Electronic and Computer Engineering, The Hong Kong University of Science and Technology, Hong Kong

Abstract: This chapter gives comprehensive description of the light outcoupling technologies for organic light-emitting diodes (OLEDs). Some state of art techniques will be introduced in detail. This chapter consists of three parts. The first part analyzes the light distribution in OLED with two models: ray optics model and dipole interference model. The second part introduces the external extraction structures, which can extract the light trapped in the substrate. The last part describes the more effective outcoupling technologies, the internal extraction structures, which can couple out not only the light in the substrate but also the light trapped in the active layers of the device. Finally, a conclusion will be given.

Keywords: Air mode, Dipole interference, External extraction structure, Internal extraction structure, Metal nanoparticles, Microlens array, Nanostructured random scattering layer, Outcoupling, Perforated hole injection layer, Photonic crystal, Ray optics, Scattering film, Substrate mode, Surface plasmon, Total internal reflection, Waveguide mode.

INTRODUCTION

Because of its great potential in flexible flat-display and solid-state lighting appli-cations, organic light-emitting diode (OLED) has attracted intensive attentions from researchers in recent years to improve the efficiency. Theore-tically, 100% internal quantum efficiency (IQE) of OLED can be achieved by applying phosphorescent emitters which make use of both the singlet and triplet excitons. In spite of the high IQE, only ~20% of the internal emission can be extracted out into the air while the remaining ~80% of the internal emission are trapped and guided in the device mainly due to total internal reflection (TIR) at the ITO/glass,

[*] **Corresponding Author Yibin Jiang:** Department of Electronic and Computer Engineering, The Hong Kong University of Science and Technology, Hong Kong; Tel.: +852-23588845; E-mail: yjiangaa@connect.ust.hk

glass/air interfaces and surface plasmons (SP) at the organic/metal interface [1].

In order to extract the 80% trapped photons, a lot of techniques have been pro-posed, including roughened substrate [1 - 3], microlens matrix [4, 5], scattering layer [6 - 10], embedded low index grids [11], high refractive index substrate [12, 13], photonic crystal structure [14, 15], and metal nanoparticles [16 - 18]. These extraction techniques can be sorted into two types: external extraction structure (EES) and internal extraction structure (IES) [8]. EES, which is set on the external surface of the substrate as its name shows, can break the TIR at the air/glass interface to extract the substrate waveguide photons. IES, which is fabricated between the device and the internal surface of substrate, is designed to interrupt the TIR at the glass/ITO interface or the SP at the organic/metal interface.

LIGHT DISTRIBUTION IN OLED

A typical OLED is just like a multilayer sandwich consisting of a planar glass substrate (d_{sub}: ~1mm, n_{sub}: ~1.5), a layer of ITO anode (d_{ITO}: ~100nm, n_{ITO}: ~1.9), one or more organic layers (d_{org}: ~100 nm, n_{org}: 1.7~1.8), and a reflecting cathode (*e.g.* Mg:Ag or LiF/Al), where d refers to the layer thickness and n refers to the refractive index. If all surfaces are planar, light emitted out of the backside of the substrate will originate only from the light at the angles less than the organic/air critical angle.

Fig. (**3.1**) shows a schematic ray diagram for a planar OLED. θ_1, given by $\sin^{-2}(n_{air}/n_{org})$, and θ_2, given by $\sin^{-2}(n_{sub}/n_{org})$, represent the critical angles at air/substrate and substrate/ITO interfaces, respectively. The ray diagram indicates that the internal emitted photons can be classified into four modes: (i) the air mode, which can escape from the substrate of the device freely ($0<\theta<\theta_1$); (ii) the substrate mode, which are trapped in the substrate by TIR at glass/air interface and thus guided to the periphery of the substrate ($\theta_1<\theta<\theta_2$); (iii) the ITO/organic mode or waveguide mode, which are confined by TIR at the ITO/substrate interface and will be quickly absorbed by ITO, organic and metal cathode layers ($\theta_2<\theta<\theta_3$); (iv) the surface plasmon (SP) mode, which are due to the coupling between the free electrons of the top metal electrode and the emission wave ($\theta_3<\theta<90°$) [1].

Assuming the cathode is a perfect reflector, and the emission is isotropic in the organic layer, the fraction of the air mode light can be calculated by

$$\eta_{air} = \int_0^{\theta_1} \sin\theta d\theta = 1 - \cos\theta_1 \approx \frac{1}{2n_{org}^2}$$

$\eta = \eta_i \times \eta_e \times \eta_{ext}$

η_i : *internal quantum efficiency*
η_e : *electrical efficiency*
η_{ext} : *light extraction efficiency*

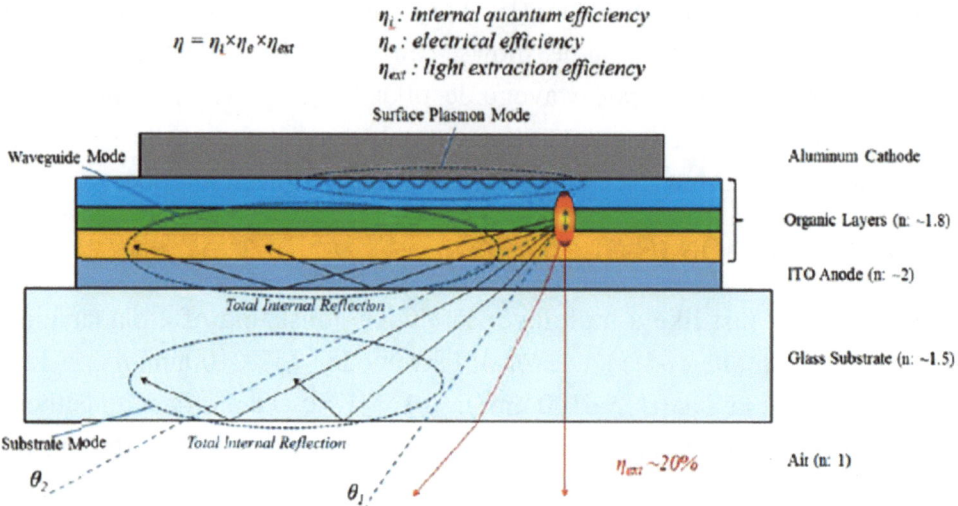

Fig. (3.1). A standard sandwich structure of OLED shows TIR at two interfaces.

The calculation indicates less than 20% light can be extracted from the device as useful lighting. The majority of the internal emitted light is trapped by TIR and waveguided inside the device until it is eventually absorbed and ultimately wasted.

The above ray optics model does not include the effects of interference of the emitted and reflected light and absorption in the various layers that make up the device. A more rigorous approach to modeling the light distribution of OLED is given below, which considers the emitters as radiating dipoles in an optical microcavity.

The general OLED structure consists of a stack of planar films arranged in the xy plane (called the plane of interfaces) as displayed in Fig. (**3.2**). The yz plane is defined as the plane of incidence. The source is considered as electric dipoles with moment of p_0 located inside an organic layer.

Therefore, the total radiation of the OLED is identical to the addition of the radiation of three independent orthogonal dipoles along the x, y, and z direction. In order to calculate the radiation dissipation into the various modes, the emitted power from dipole sources is represented as an integral which is calculated from zero to infinity over the magnitude of the inplane wavevector $k_{//}$.

$$P = P^{(s)} + P^{(p)} = \int_0^\infty [W^{(s)}(k_{//}) + W^{(p)}(k_{//})]dk_{//}$$

The calculation of the power density $W^{(s,p)}(k_{//})$ of an oscillating dipole embedded in a multilayer structure can be carried out using the transfer matrix method and multilayer stack theory. Fig. (**3.3**) shows the $k_{//}$ space power spectrum assuming randomly oriented dipoles. Wavevector k_0 refers to the peak emission wavelength of the dipole in vacuum condition, while k_{org} to that when the dipole is embedded in organic layers [19]. The power fractions of different modes can be represented by the area under the power density curve in the corresponding mode's wavevector $k_{//}$ range. Because of the interfaces between different materials, the total radiation power is dissipated into different modes. The power with inplane wavevectors that meet the condition of $0<k_{//}<k_{glass}=n_{glass}k_0$ corresponds to light that may transmit into the glass substrate. Of this dissipated power, the fraction with $k_{//}$ in the range of $0<k_{//}<k_0$ corresponds to the air mode capable of producing useful lighting, while the fraction with $k_0<k_{//}< k_{glass}$ is the substrate mode. The fraction with $k_{//}$ between k_{glass} and k_{org} corresponds to the waveguide mode in the ITO and organic layers. There exists a sharp peak when $k_{//}$ is a little larger than k_{org}, which represents the SP mode. Since the wavevectors of the SP mode are larger than k_0, the SP is strongly confined to the interface between the metal electrode and the organic layer, thus the SP mode cannot escape into the air.

To quantify the fraction of power coupled to a specific mode, the area under the curve associated with that mode is integrated and normalized to the total area under the power spectrum. For a thickness optimized device, the dipole interference model predicts that the fractions of air mode, substrate mode, waveguide mode and SP mode are 29%, 29%, 12% and 30% respectively. The new fraction value of air mode is 55% larger than that predicted by ray optics model. It infers that the dipole radiation in the emitting layer is not isotropic. The interference effect from the reflective cathode strongly modifies its emission, enhancing the radiation in normal direction while suppressing that at large angle.

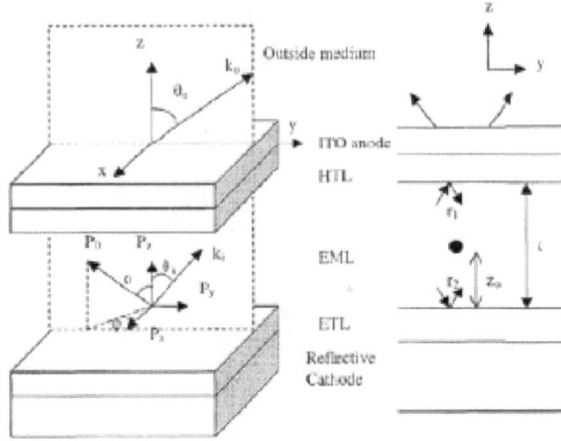

Fig. (3.2). General structure of multilayer organic light-emitting devices, 3D (left) and 2D (right).

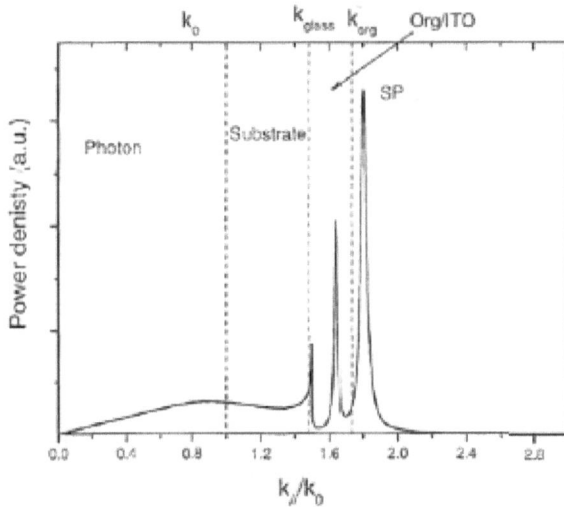

Fig. (3.3). Calculated power dissipation spectrum of the conventional OLED.

EXTERNAL EXTRACTION STRUCTURES

From the above model simulation, there are considerable photons trapped in the substrate. Therefore, substrate modification is an effective way to increase the outcoupling of OLED without influencing the active layer of the device. These modifications are called external extraction structures (EESs) since they are built outside the device structure and can access only to substrate mode. A few representative EESs will be introduced and analyzed in the following part.

Truncated Square-pyramid Luminaire

This luminaire was invited by Universal Display Corporation [20], as exhibited in Fig. (**3.4**). The luminaire is made from acrylic which has a refractive index of 1.5. The dimensions of the luminaire are indicated in the figure with base angles θ of 30°, 45°, and 60°, base side length of 7.1 cm, and height of 1.2 cm. The luminaire was optically coupled to the OLED glass substrate using index matching fluid having n=1.5. The forward emission of an OLED with a truncated square-pyramid luminaire can be improved to be 1.8–2.1 times greater than that of an OLED without a luminaire [20].

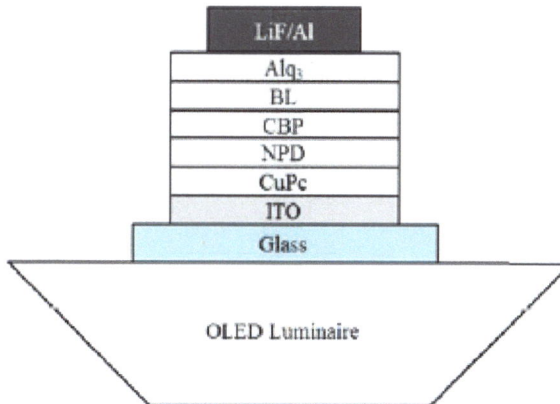

Fig. (3.4). A schematic diagram of an OLED with a luminaire. The clear acrylic luminaires are shaped to be a truncated square pyramid with various base angles θ.

Scattering Film

The above luminaire seems too thick compared to the device structure of OLED. It will be awkward when this luminaire is applied in flat-panel display. External scattering film is another solution to disturb the TIR at the glass/air interface and extract the substrate mode, while still keeping the whole device structure thin. To make scattering films, there are many fabrication methods, which can be classified into two groups: solution process method and vacuum process method. For solution process, the fabrication steps include (1) mixing micro or nano particles such as silica [21], poly-microspheres [22], ZrO_2 [23, 24], TiO_2 [8], or ZnO [25] with the host solution like PDMS (poly dimethyl silicone resin) or transparent photoresist, (2) spin-coating the mixture on the substrate and (3) curing the film, as shown in Fig. (**3.5**). The particles can redirect the paths of the trapped light and finally extract the substrate mode out, as illustrated in Fig. (**3.6**). Large particle size and high concentration of the particles can enhance the film's scattering capability. The outcoupling performance of these scattering films is summarized in Table (**3.1**).

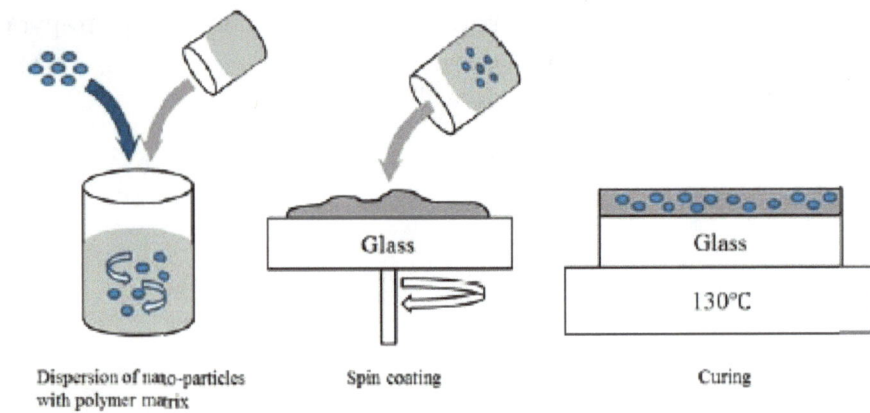

Dispersion of nano-particles with polymer matrix   Spin coating   Curing

Fig. (3.5). Procedures of preparing the micro or nano particles embedded scattering film.

Table 3.1. Summary of Different External Scattering Films.

Scattering structure	Efficiency enhancement	Ref
poly-microsphere in acrylate-based resist	22%	22
ZrO_2 microparticles in PDMS	40%	23,24
TiO_2 nanoparticles in transparent photoresist	60%	8
ZnO nanoparticles in polystyrene	30%	25

Although the particles embedded scattering films can effectively couple out the substrate mode, the fabrication processes may be somewhat complex with so many steps. It is also hard to achieve uniform film as the particles tend to aggregate during the whole fabrication process. Besides, the solution fabrication process sets a limit for this technique to be applied as an internal or external scattering layer in the top-emitting OLEDs. These disadvantages can be overcome by vacuum based processes. Chen *et al* developed two vacuum based fabrication methods to grow scattering films [6 - 7]. The two methods make use of the property that crystallization of small molecular organic materials usually results in nanostructures with rough surface. In one method, vacuum deposited amorphous 4,40-bis(1,2,2-triphenylvinyl)biphenyl (BTPE) film can be converted to nanowires scattering film after annealing at a temperature of 110°C. Fig. (**3.7**) shows the scanning electron microscopy (SEM) picture of the BTPE nanowires film together with its haze measurement results. The haze value, which is defined as $HAZE = (T_{total} - T_{specular}) / T_{total}$, is used to indicate what percentage the scattered light takes up and characterized the light scattering ability of the film. With this scattering film, the device performance in regarding of the efficiency and luminance is greatly enhanced because of the strong light scattering, as indicated by the curves in Fig. (**3.8**). The best efficiency enhancement is 31.5% with the 300nm BTPE nanowires scattering film.

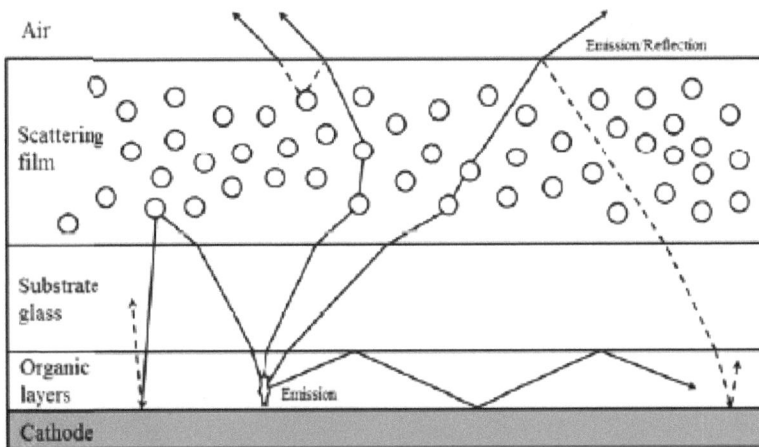

Fig. (3.6). The light ray behaviors in an OLED with external scattering film.

Another method is much simpler with just one-step fabrication. In this work, nano-particles scattering film was facilely fabricated by just one-step vacuum evaporation of an organic compound tetraphenylethene (TPE), which has a very low Tg. Usually, researchers tend to avoid low Tg materials as those materials form nanocrystal easily at low temperature and thus the interface of film is rough, which is harm to the device stability and lifetime. Here, this "adverse" effect was utilized to explore its application in serving as light scattering film for OLEDs. It was found that the TPE molecules crystallize themselves immediately when the molecules are evaporated onto the substrate [7]. When the TPE film is magnified under SEM, the assembly of the TPE molecules shown in the form of microparticles as shown in Fig. (3.9). These microparticles can efficiently scatter the light. As a result, a 28% efficiency enhancement can be obtained by applying the as-deposited TPE microparticles as the scattering media for OLEDs.

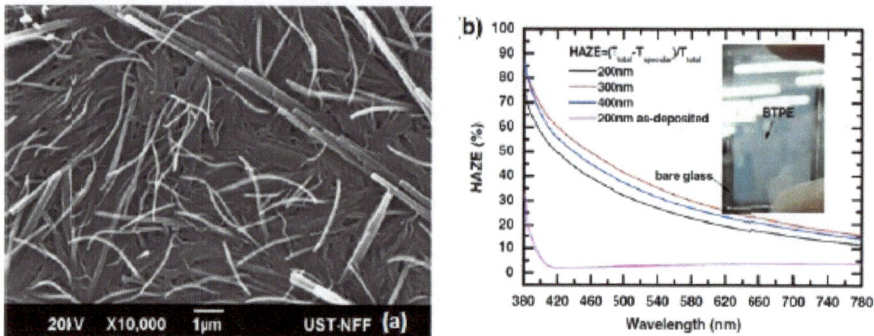

Fig. (3.7). (**a**) Top SEM view and (**b**) haze spectrum of the annealed BTPE film on glass substrate. Inset of (**b**) shows the photo of the annealed BTPE film. Due to strong light scattering effect, the film exhibits a milky appearance.

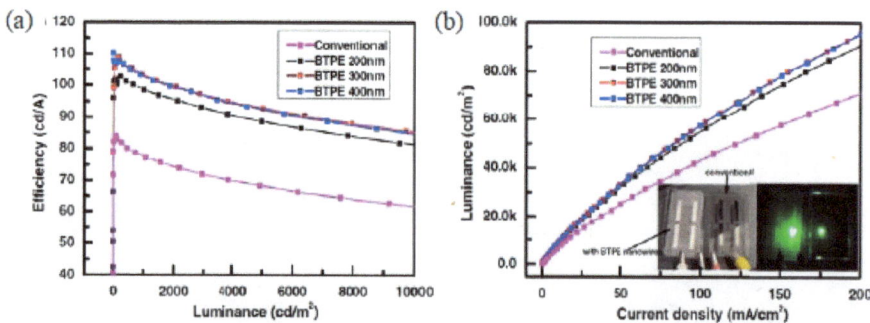

Fig. (3.8). (**a**) Efficiency-luminance and (**b**) luminance-current density characteristics of the BTPE nanowires scattering film coated OLEDs. Inset of (**b**) shows the photos of the OLEDs. The nanowires modified OLED emits brighter than the conventional OLED.

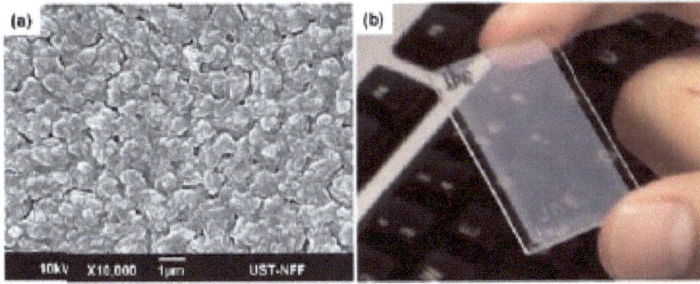

Fig. (3.9). (**a**) SEM image of the 300nm as-deposited TPE film, and (**b**) Photo of the 300nm as-deposited TPE scattering film.

Sand-blasting Substrate

Although light extraction by employing scattering film is an effective choice for large-area display or lighting application, it increases the fabrication cost and complexity. Therefore, roughening the surface of the glass substrate directly is a relatively simple and cost-effective way to extract the substrate mode. Chen proposed using sand-blast to roughen the substrate as scattering medium [26]. By blasting the sand particles to the edges and back of the glass substrate to roughen the surface, the forward efficiency of the device can achieve a 20% improvement, as exhibited by the efficiency-current density characteristic in Fig. (**3.10**).

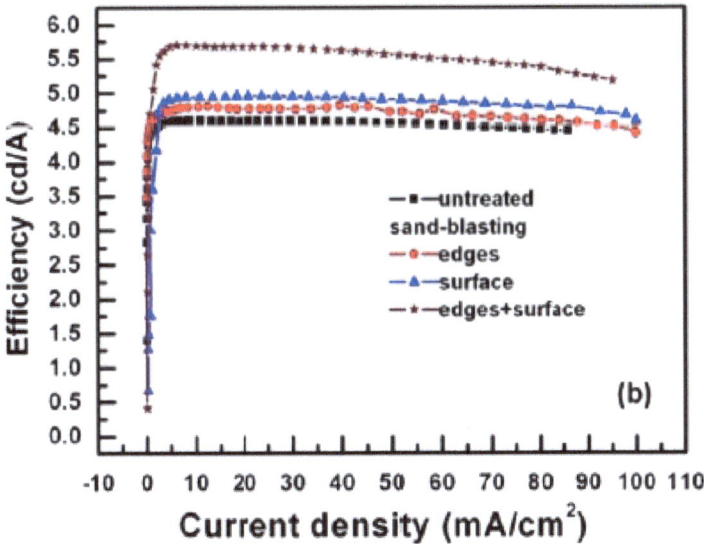

Fig. (3.10). Efficiency-current density characteristic of the OLEDs fabricated on sand-blasted substrates.

The microscope images of the sand-blasted glass substrate with different magnifications are shown in Fig. (3.11) together with the optical model illustration. The close observation under the microscope reveals that the rough surface consists of a series of random distributed grains with an average size of ~100um, which is just the size of the sand particles [26]. These grains can disturb the TIR at the glass/air interface and thus increasing external efficiency. According to the illustration in Fig. (3.11b), by sand-blasting the back of the substrates, the original wave-guided light can be scattered out as useful light, as the comparison between ray b and b'. In addition, by sand-blasting the edges of the substrates, rays with large incidence angle which originally escape from the edges (rays c, d) may be scattered into forward emission by the rough edges (ray c'), or be scattered back to the substrate and may have the chances to escape from the rough back surface (ray d').

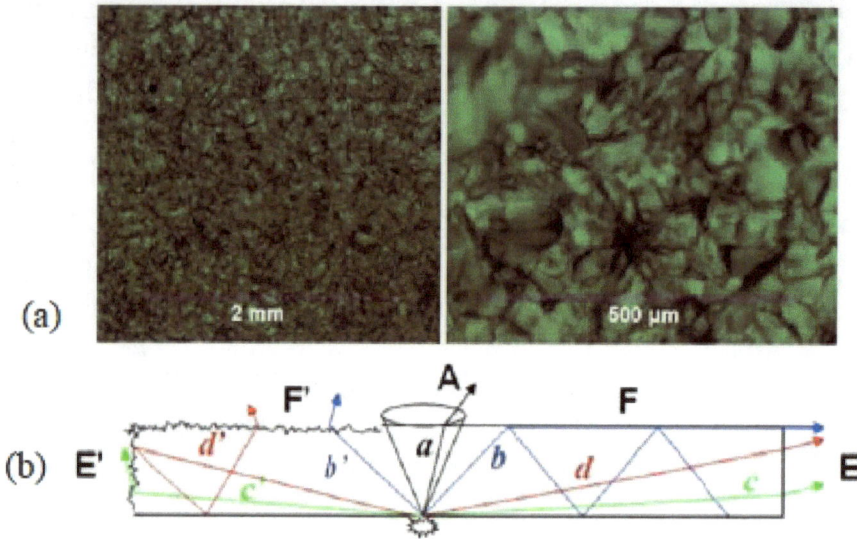

Fig. (3.11). (a) Microscope images of the glass substrates after sand-blasting. (b) Schematic illustration of light propagation inside the sand-blasted substrate.

Microlens Array

The outcoupling mechanism of microlens is to decrease the incidence angle of the TIR light to below the critical angle, as illustrated in Fig. (3.12). The fabrication of microlens array involves mold transfer process which is a large-area process. Fig. (3.13) shows the fabrication steps of one kind microlens array together with its SEM picture [4]. First, a 1 μm thick SiN_x layer is grown on silicon wafer by plasma enhanced chemical vapor deposition (PECVD). Then a square hole pattern is formed in the SiN_x layer using standard photolithography (step 1) plus etching process.

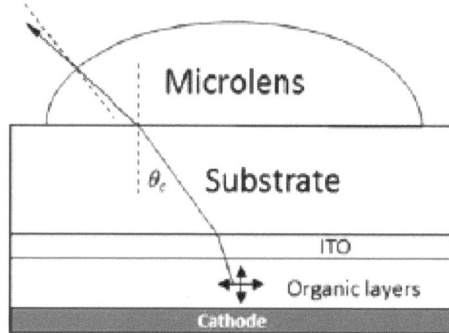

Fig. (3.12). The light path in an OLED with microlens. The incidence angle of the TIR light can be reduced.

After removing the photoresist (step 2), the Si wafer is immersed into the HNO_3: CH_3COOH:HF (8:1:1) solution to etch silicon using the SiN_x layer as the mask (step 3). After that, the SiN_x is removed by etching in HF to get the silicon mold (step 4). Since there exists undercut under the SiN_x layer due to isotropic etching of silicon, the size & shape of the starting pattern and the etching time will determine the final shape of the mold. At last, the mold is subsequently filled with PDMS, a thermally curable elastomer, and then stuck to glass substrate, cured and peeled off to obtain the microlens array sheet. With this microlens sheet attached to glass substrate, the light output of OLED can be increased by a factor of 1.5 over unlensed substrate.

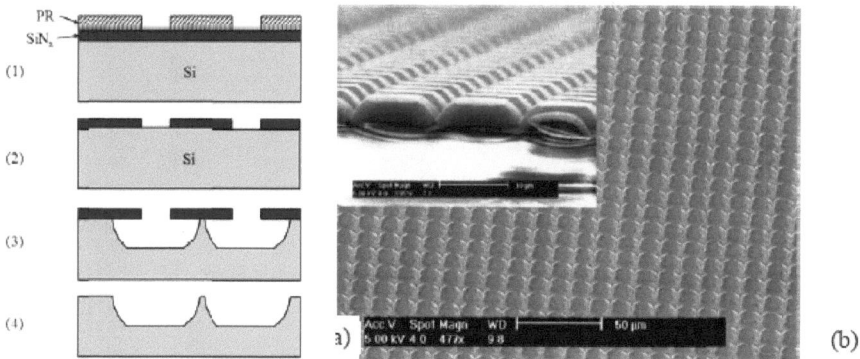

Fig. (3.13). (a) Fabrication process of the silicon mold. **(b)** SEM of a PDMS microlens array fabricated from this silicon mold. Inset of (b) shows the detailed side view of the PDMS microlens which accurately images the mold shape.

However, it can be observed that the shape of the lens is trapezoid rather than sphere. Peng develop a new fabrication method to achieve spherical shape microlens array as schematically shown in Fig. (**3.14**) [5]. The first step of the fabrication is to spin-coat a layer of photoresist on glass. Second, disk arrays are patterned by conventional photolithography. The photoresist disks are then heated to the melting point and reflowed to form spherical shape. After that, the negative pattern of the photoresist semi-sphere is replicated to PDMS to form the mother mold of the microlens array. Finally, this PDMS mold with epoxy resin filling is attached to the backside of the glass substrate. After a few minutes exposure under UV light, the epoxy is cured to be a transparent hard film with microlens array on it. With this spherical microlens array, a larger improvement factor of 1.7 was obtained experimentally compared to the above method, indicating the importance of the shape of the lens.

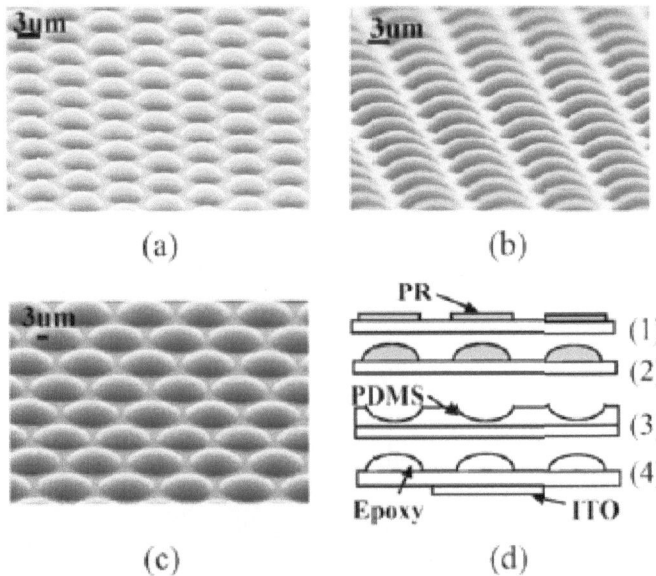

Fig. (3.14). SEM images of the microlens arrays with diameter of: (**a**) 5μm; (**b**) 10μm; (**c**) 15μm. The spacing between the lenses is 1μm for all cases. (**d**) Schematic illustration of the lens fabrication process.

INTERNAL EXTRACTION STRUCTURES

Since the EES has access only to the substrate mode, its outcoupling performance is quite limited. Thus, people tried to fabricate the extraction structures under or in the active layers of the device to couple out not only the substrate mode but also the waveguide mode. As a result, these structures are known as internal extraction structures (IESs). Although the performance of IES is very impressive, the fabrication of IES faces more challenges due to the risk of potential short circuit problem caused by the rough surface of IES. The following part will describe a few IESs including internal scatter layer, photonic crystal structure and metal nanoparticles.

Internal Scattering Layer

The working principle of the internal scattering layer is just same as that of the external scattering film. Conventionally, this layer is just below and in intimate contact with the bottom electrode of the device. Nowadays, people have successfully combined this internal scattering layer with the active layers of the device. The conventional internal scattering layer should be talked firstly. The whole fabrication process involves three main steps: (1) form scattering structure on the substrate, (2) planarize the scattering surface, and (3) deposit the active layers of the device. Fig. (**3.15**) describes the fabrication process of a nanostructured random scattering layer (RSL). This method makes use of the dewetted Ag droplets as a hard mask to texture the glass surface as random scattering structure [27]. It should be noted that a planarization layer is coated on the RSL. Planarization is necessary to stabilize the forthcoming deposition of the active layers and suppress any anomalies in the current flow characteristic.

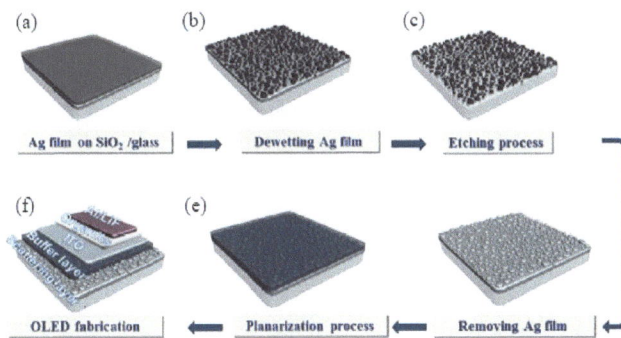

Fig. (3.15). The schematic of process flow for preparing the RSL. (**a**) Preparation of the substrate, (**b**) formation of irregularly patterned Ag mask by a dewetting process, (**c**) dry etching process, (**d**) removal of metal mask for forming the scattering layer, (**e**) planarization process, and (**f**) OLED fabrication.

Fig. (**3.16**) shows another fabrication scheme of an internal scattering layer which is a SiO_2-microsphere monolayer [28]. Monodisperse SiO_2-microparticles solution is spin-coated on glass in an acceleration manner with speed ramping up to 2000 rpm at 500 rpm/s to generate a monolayer. Then a ~1µm thick layer of photoresist SU-8 is spin-coated on the SiO_2 spheres in order to fill the spaces between the spheres. This is necessary to guarantee continuous film of the following deposited layers.

Although the above internal scattering layers were effective to extract the waveguide mode, the fabrication is somewhat complex. Therefore, Chen's group developed an IES based on nanostructured aluminum doped zinc oxide (AZO) that combine the electrode and the scattering layer in one single entity, with simple wet-chemical etching process [29]. Because of the AZO's polycrystal feature, nanostructure can be easily formed on the surface of the film after wet-etching process. The SEM images of the nanostructured AZO samples after different etching time are shown in Fig. (**3.17**) together with the device structure. The nanostructured AZO contains protruding features, which may cause electrical instability during the device operation. For the purpose of preventing the short circuit, thick hole-transport layer HATCN with a thickness of 60nm is used.

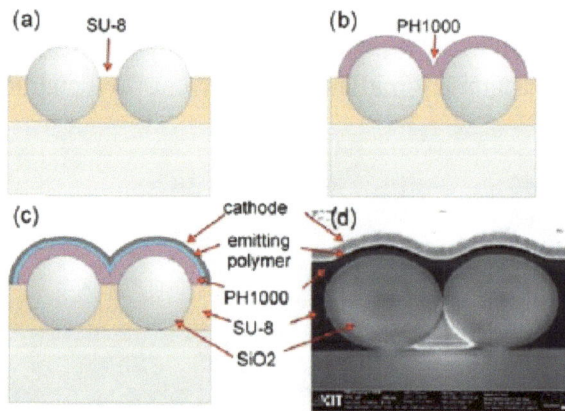

Fig. (3.16). Fabrication scheme. (**a**) A layer of SU-8 was spin-coated on the sphere-covered substrates to fill the spaces between the spheres. (**b**) A layer of PEDOT:PSS was spin-coated on top to serve as the anode of the OLED. (**c**) A polymer layer and LiF/Al layer were sequentially deposited to serve as the emitting layer and the cathode, respectively. (**d**) SEM-image of a cross section of the finished device.

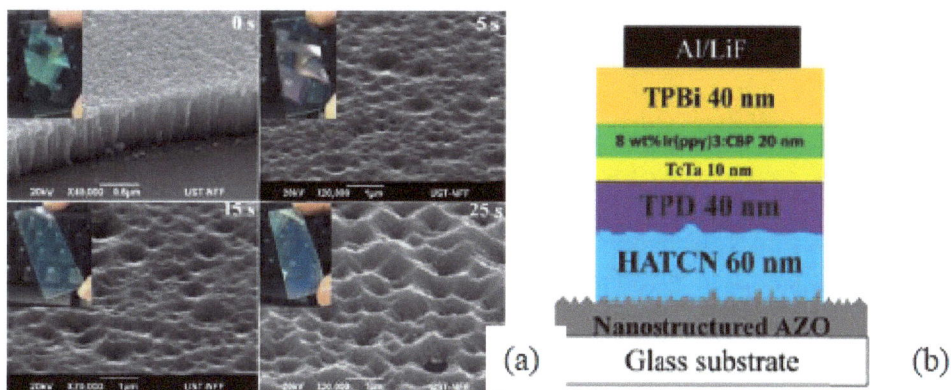

Fig. (3.17). (**a**) Scanning electron microscopy (SEM) images and photos of the AZO samples after 0 s, 5 s, 15 s, and 25 s etching. (**b**) Structure of OLED fabricated on nanostructured AZO substrate.

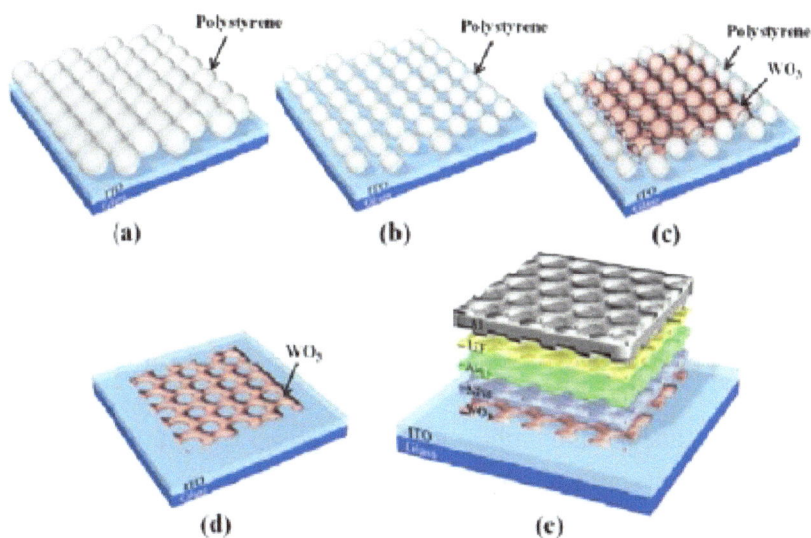

Fig. (3.18). Graphical illustration of the fabrication process of the OLED device with the perforated WO^3 layer as HIL: (**a**) a well ordered PS monolayer was formed by lifting up the substrate from a solution with PS nanosphere floating on the surface, (**b**) the closely packed PS monolayer changed to a non-closely packed self-assembled PS monolayer after the plasma etching, (**c**) a WO^3 layer was deposited on the PS monolayer, (**d**) after lift-off process using sonication a perforated WO^3 layer was obtained, and (**e**) finally different functional layers of the OLED were evaporated onto the perforated WO^3 layer sequentially.

The internal scattering layer can also be combined into the hole injection layer (HIL). As shown in Fig. (**3.18**), people use polystyrene (PS) nanosphere monolayer as mask to fabricate a perforated WO_3 hole injection layer [30]. With this nanostructured WO_3 HIL, the light extraction efficiency can be enhanced because of Bragg scattering of waveguide modes.

The fabrication and key characteristics of the representative internal scattering layers are summarized in Table (**3.2**).

Table 3.2. Summary of different internal scattering layers.

Structure	Fabrication	Efficiency enhancement	Ref
Nanostructured random scattering layer (RSL)	Ag droplets mask + Dry etching	52%	[27]
TiO_2 nanoparticles in polymer host	Spin coating + Curing	96%	[8]
Nanosized random texture layer (nRTL)	Reactive ion etching + Printing transfer	61%	[31]
SiO_2-microspheres monolayer	Spin coating	200%	[28]
Nanostructured AZO	Wet-chemical etching	60%	[29]
WO_3 nanoislands	Ag droplets mask + Wet-chemical etching	45%	[32]
Perforated WO_3 layer	polystyrene nanosphere monolayer as mask + deposition	53%	[30]

Photonic Crystal Structure

Previously, two-dimensional photonic crystal (PC) patterns have been incorporated into thin slab InGaAs light-emitting diode in order to improve the external efficiency [33]. Similarly, PC structure can be applied in OLED device. The optimization of PC's structural parameters is very important, since the light extraction will become the most efficient when the lattice constant of the PC structure gets close to the vacuum wavelength as predicted by simulation [14]. Fig. (**3.19**) shows the configuration of a PC-OLED. The fabrication process is described as follows. First, a 200nm thick SiO_2 layer is grown on the glass substrate. Then, after irradiated holographic lithography followed by reactive ion etching, the PC pattern with 200nm depth are formed on the SiO_2 layer. After that, thick SiN$_x$ layer is deposited on the SiO$_2$ PC pattern by PECVD. The total thickness of the SiO$_2$/SiN$_x$ PC structure is 800nm, giving a nearly flat top surface which guarantees the good surface quality of the subsequently deposited active layers of the device. With this PC, an over 50% improvement of the external efficiency was obtained.

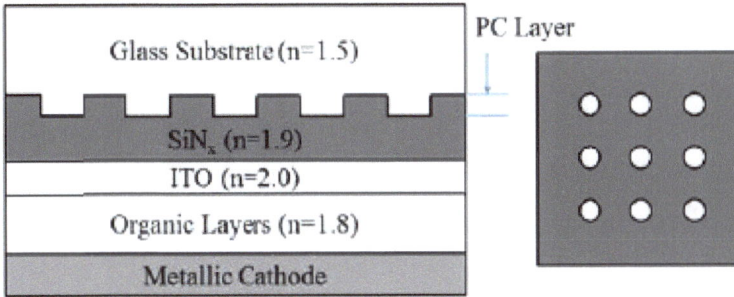

Fig. (3.19). The layer structure of a PC-OLED.

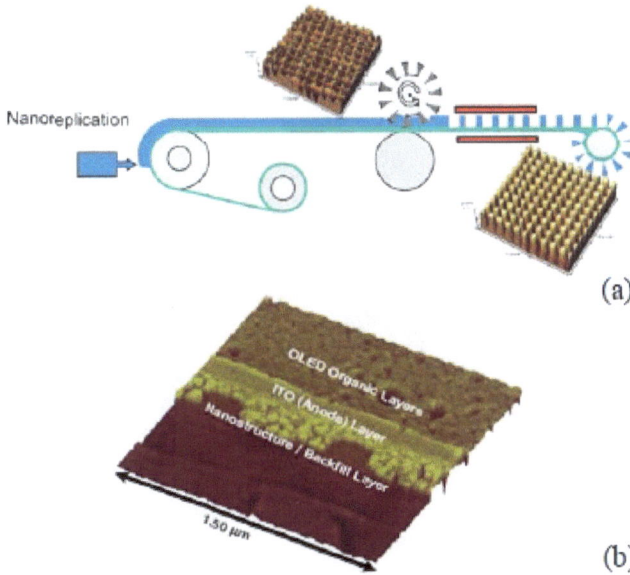

Fig. (3.20). (a) Roll-to-roll replication system for fabricating PC pattern. (b) Cross-sectional AFM image of a bottom emitting OLED on a PC nanostructure replica substrate.

However, the above fabrication involves holographic lithography which is small-area technology. In order to overcome this limitation, people have adapted roll-to-roll replication technology to fabricate large area PC pattern [34]. Fig. (**3.20a**) illustrates a nanoreplication system. Substrate coated with curable polymer goes through a casting roll which is equipped with a nano-scale mold surface. After passing, the nano-pattern of the mold will be transferred onto the curable polymer. Then the replicated pattern is solidified by curing tool.

Because this system can continuously handle the substrate with roll tool, it is possible to efficiently and accurately produce PC pattern on large area substrates. The insets within Fig. (**3.20b**) provide surface AFM graphs comparison between the mold and the replica, which exhibits great duplication ability. Using this nanoimprinted PC, a 70% efficiency enhancement can achieved.

Metal Nanoparticles

According to the above dipole interference model, 30% of the generated photons are lost to surface plasmon (SP) mode at the metal/organic interface. The excitation of the surface plasmons at the metal contact guides the light along the smooth metal surface in the OLEDs device, thus quenching the emission. Localized surface plasma (LSP) can disturb this lateral progress of SP and help improve the overall efficiency. Fortunately, metal nanoparticles (NPs) can be used to generate LSP. Up to now, Au NPs, Ag NPs and Pt_3Co alloy NPs have been introduced into the OLED device to extract the SP mode.

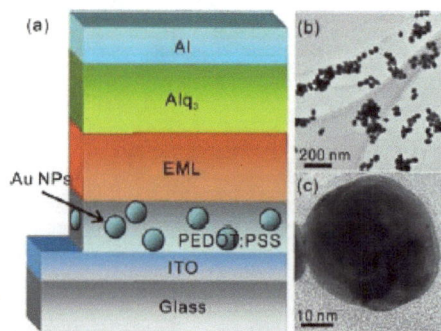

Fig. (3.21). (**a**) Schematic device structure of an OLED incorporating Au NPs in the PEDOT:PSS layer. (**b**) TEM and (**c**) high-resolution TEM images of the synthesized Au NPs.

Au NPs are prepared through the standard sodium citrate reduction method as reported previously [35]. Au NPs are dispersed into the aqueous solution of PEDOT:PSS at an optimized ratio. The resulting solution is spin-coated onto UV-zone treated ITO glass substrate and subsequently annealed at 110 °C in air. The device structure as well as TEM of Au NPs is shown in Fig. (**3.21**). With the help of Au NPs, an increase of ~25% in the EL intensity and efficiency are achieved, whereas the spectral and electrical properties remain almost identical to the control device without Au NPs.

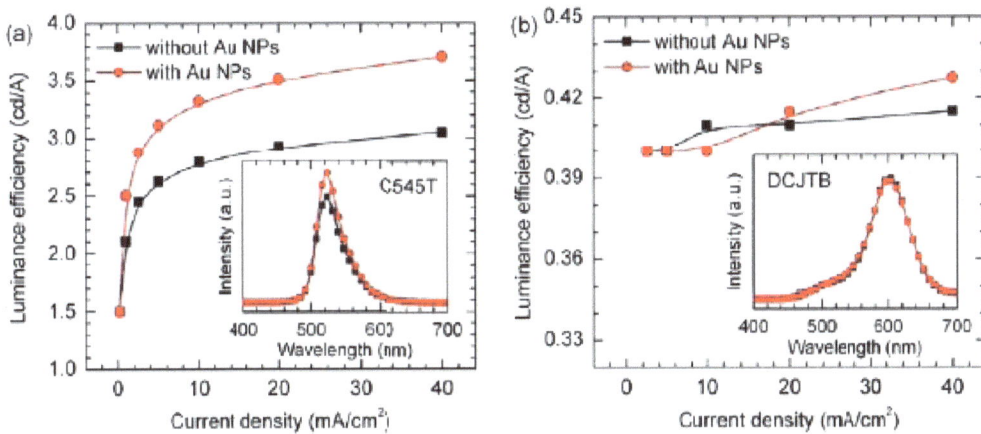

Fig. (3.22). Comparison of current efficiency of (**a**) green emission and (**b**) red emission OLEDs with and without Au NPs in the PEDOT:PSS layer. Insets are the corresponding EL spectra at a current density of 20mA/cm^2.

To further clarify the effect of localized SP resonance on the EL enhancement, the OLEDs with various emitters were fabricated, in which C545T and DCJTB were used as dopants for green and red emissions, respectively. As shown in Fig. (**3.22a**), there is a ~20% improvement for the C545T device. Meanwhile, the inset of Fig. (**3.22a**) shows a significant EL enhancement for green emission without any spectral change. This is because the resonance energy of the LSP excitation mode matches with the excitons of C545T in OLEDs, leading to an effective LSP enhancement by Au NPs. On the contrary, there is no observable EL enhancement of the OLED with red emission as shown in Fig. (**3.22b**). As the EL peak of DCJTB is located at 600 nm, there is lack of the overlap between the emission of DCJTB and the resonance energy of the SP excitation mode. Therefore, it is expected that the coupling between radiated red emission of DCJTB and LSPs of Au NPs is blocked.

Ag NPs can be formed by evaporating a very thin Ag layer [17]. For example, when the monitored thickness is 1 nm, a non-continual Ag sublayer is actually formed on the substrate as shown in Fig. (**3.23b**). Thus, the Ag NPs exist in the form of cluster with diameter ranging from 1 to 13 nm. This Ag NPs layer can be inserted into the organic layer to induce LSP. But the position of this Ag sublayer needs to be finely tuned during the evaporation process in order to achieve the maximum coupling between the LSP and the exciton. The device structure is shown in Fig. (**3.23a**) with Ag NPs embedded in the electron transportation layer.

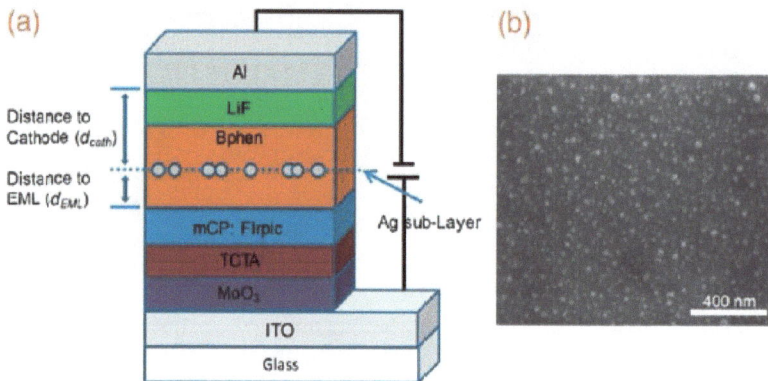

Fig. (3.23). (**a**)Schematic diagram of an OLED device with a thin Ag NPs layer inserted in electron transportation layer, and (**b**) SEM image of 1 nm thick Ag sublayer on a substrate.

The relative position of the Ag NPs layer, *i.e.*, distance to EML (d_{EML}) and cathode (d_{cath}), was varied to investigate the optimum condition. Fig. (**3.24**) shows the change of the devices' current density-voltage-current efficiency (J-V-η) characteristic with different Ag NPs layer positions in the 40nm Bphen layer. According to the performance, it can be found that when the Ag NPs layer is very close to the EML exciton quenching occurs easily, and that an EL enhancement of 18% can be obtained due to optimized energy transfer when the Ag NPs layer is set at the middle of the electron transportation layer.

Except the single element metal nanoparticles, alloy nanoparticles (ANPs) can also be synthesized in order to combine the features of the different elements. As a result, Pt_3Co ANPs were synthesized because the Pt and Co elements are more inert than the normal metals (such as Au and Ag). Thus, the ANPs will not aggregate easily and endure the following thermal treatment. Besides, the Pt_3Co ANPs can directly contact the anode surface as the workfunction of the ANP is just between that of ITO and the HOMO level of hole transportation materials.

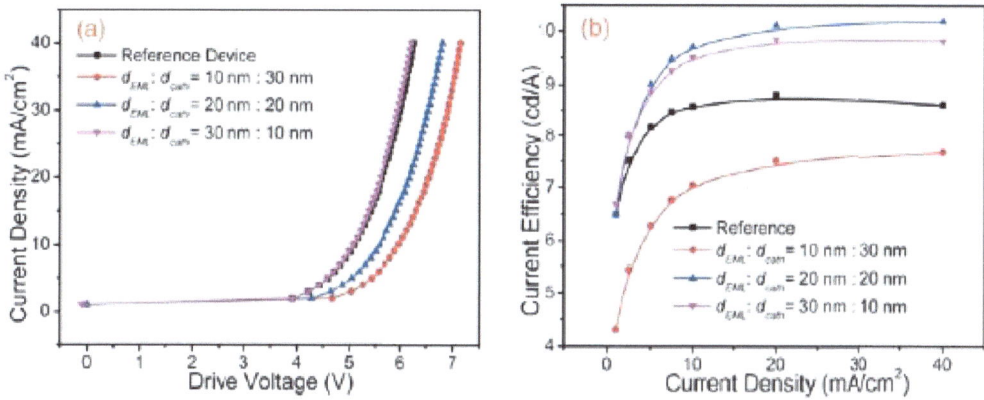

Fig. (3.24). Performance comparison of the devices with Ag layer position as a control parameter in regarding of (**a**) current density-voltage and (**b**) current efficiency-current density characteristics. The reference is a standard device without Ag in it.

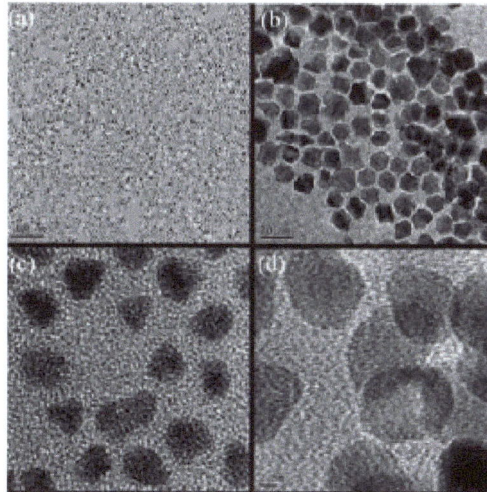

Fig. (3.25). (**a-d**) TEM images of the synthesized Pt3Co ANPs with different magnifications.

Monodisperse Pt_3Co ANPs are synthesized through an organic solvothermal approach modified from previous publications [36], which consists of the following two steps: (1) synthesis of the C18PMH-PEG polymer; (2) PEG functionalization of Pt_3Co ANPs.

Fig. (3.25) show the high resolution TEM images of the synthesized Pt_3Co ANPs. It can be found from these figures that Pt_3Co ANPs are not perfect spherical with many edges and corners in shape and have a small size distribution with an average diameter of ~8 nm.

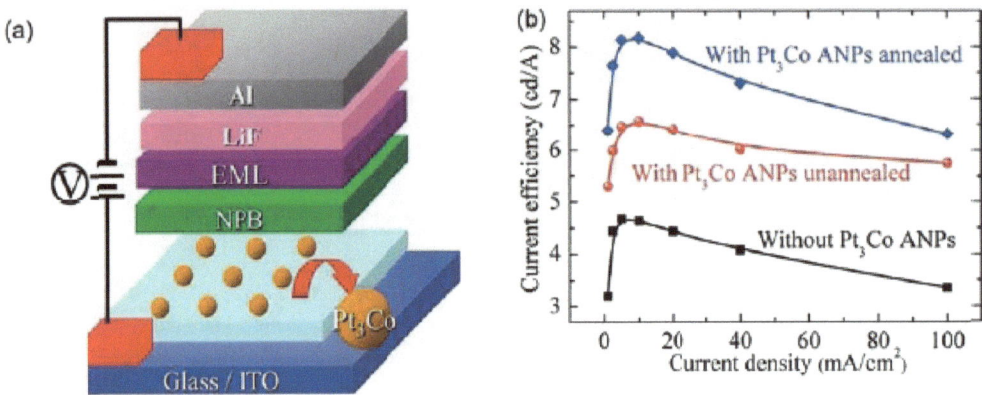

Fig. (3.26). (**a**) Device layer structure of an OLED incorporating Pt3Co ANPs on ITO, (**b**) efficiency-current density characteristic of three kinds Alq3 based OLEDs: device A (red line), device B (blue line) and device C (black line).

Pt_3Co ANPs can be applied in the device as an anode buffer layer. Fig. (**3.26a**) shows a schematic diagram of the device layer structure with Pt_3Co ANPs on ITO. Three devices are investigated (Device A: ITO/unannealed Pt_3Co/NPB 40nm/EML 50nm/LiF 1nm/Al 100nm; Device B: ITO/annealed Pt_3Co/NPB 40nm/EML 50nm/LiF 1nm/Al 100nm; Device C: ITO/NPB 40nm/EML 50nm/LiF 1nm/Al 100nm). The performances of device A, B and C are compared in Fig. (**3-26b**). According to the figure, the device incorporating unannealed Pt_3Co ANPs shows a 41% efficiency enhancement at a driving current density of 10 mA/cm^2 as a result of the LSP and the enhanced near-field which are induced by the ANPs. Moreover, the following annealing process has an additional effect on the device efficiency. The current efficiency is further enhanced by 26%

compared to that of the device without thermal treatment. This further enhancement is attributed to the aggregation of some Pt_3Co ANPs during the annealing process, which can cause light scattering and boost the light extraction.

CONCLUSION

Different light outcoupling technologies have been reviewed in this chapter, including external EESs and IESs. EES can extract the substrate mode by disturbing the TIR at the air/glass interface, using luminaire, scattering film, sandblasted substrate and microlens array. Comparing to EES, the fabrication of IES is more complex and challenging in order to ensure the electrical reliability of the device. But IES can couple out not only the substrate mode but also the waveguide mode and SP mode, thus more effective than the EES. In this chapter, the state of art IESs, including internal scatter layer, photonic crystal structure and metal nanoparticles, are described in detail. The fabrications of these IESs involve many nanotechnologies, such as Ag droplets mask, nanoimprint and nanoparticle synthesis. If these outcoupling technologies mentioned above can be incorporated into real applications to improve the efficiency, the battery lifetime of the devices using OLED components will be dramatically increased as the display module is one of the most power-consuming parts.

CONFLICT OF INTEREST

The authors confirm that this chapter contents have no conflict of interest.

ACKNOWLEDGEMENTS

Declared none.

REFERENCES

[1]   S. Reineke, F. Lindner, G. Schwartz, N. Seidler, K. Walzer, B. Lüssem, and K. Leo, "White organic light-emitting diodes with fluorescent tube efficiency", *Nature,* vol. 459, no. 7244, pp. 234-238, 2009. [http://dx.doi.org/10.1038/nature08003] [PMID: 19444212]

[2]   H.J. Peng, Y.L. Ho, X.J. Yu, and H.S. Kwok, "Enhanced coupling of light from organic light emitting diodes using nanoporous films", *J. Appl. Phys.,* vol. 96, pp. 1649-1654, 2004. [http://dx.doi.org/10.1063/1.1765859]

[3]   Y-H. Cheng, J-L. Wu, C-H. Cheng, K-C. Syao, and M-C. Lee, "Enhanced light outcoupling in a thin film by texturing meshed surfaces", *Appl. Phys. Lett.,* vol. 90, p. 091102, 2007.

[http://dx.doi.org/10.1063/1.2709920]

[4] S. Möller, and S.R. Forrest, "Improved light out-coupling in organic light emitting diodes employing ordered microlens arrays", *J. Appl. Phys.,* vol. 91, pp. 3324-3327, 2002.
[http://dx.doi.org/10.1063/1.1435422]

[5] H.J. Peng, Y.L. Ho, X-J. Yu, M. Wong, and H-S. Kwok, "Coupling efficiency enhancement in organic light-emitting devices using microlens array-theory and experiment", *J. Displ. Technol.,* vol. 1, pp. 278-282, 2005.
[http://dx.doi.org/10.1109/JDT.2005.858944]

[6] S. Chen, Z. Zhao, B.Z. Tang, and H-S. Kwok, "Growth methods, enhanced photoluminescence, high hydrophobicity and light scattering of 4,40-bis(1,2,2-triphenylvinyl)biphenyl nanowires", *Org. Electron.,* vol. 13, pp. 1996-2002, 2012.
[http://dx.doi.org/10.1016/j.orgel.2012.06.014]

[7] S. Chen, W. Qin, Z. Zhao, B.Z. Tang, and H-S. Kwok, "One-step fabrication of organic nanoparticles as scattering media for extracting substrate waveguide light from organic light-emitting diodes", *J. Mater. Chem.,* vol. 22, pp. 11386-11390, 2012.

[8] H-W. Chang, K-C. Tien, M-H. Hsu, Y-H. Huang, M-S. Lin, C-H. Tsai, Y-T. Tsai, and C-C. Wu, "Organic light-emitting devices integrated with internal scattering layers for enhancing optical out-coupling", *J. Soc. Inf. Disp.,* vol. 19, pp. 196-204, 2011.
[http://dx.doi.org/10.1889/JSID19.2.196]

[9] B. Riedel, J. Hauss, M. Aichholz, A. Gall, U. Lemmer, and M. Gerken, "Polymer light emitting diodes containing nanoparticle clusters for improved efficiency", *Org. Electron.,* vol. 11, pp. 1172-1175, 2010.
[http://dx.doi.org/10.1016/j.orgel.2010.04.017]

[10] W.H. Koo, S.M. Jeong, F. Araoka, K. Ishikawa, S. Nishimura, T. Totyooka, and H. Takezoe, "Light extraction from organic light-emitting diodes enhanced by spontaneously formed buckles", *Nat. Photonics,* vol. 4, pp. 222-226, 2010.
[http://dx.doi.org/10.1038/nphoton.2010.7]

[11] Y. Sun, and S.R. Forrest, "Enhanced light out-coupling of organic light-emitting devices using embedded low-index grids", *Nat. Photonics,* vol. 2, pp. 483-487, 2008.
[http://dx.doi.org/10.1038/nphoton.2008.132]

[12] M-H. Lu, and J.C. Sturm, "Optimization of external coupling and light emission in organic light-emitting devices: modeling and experiment", *J. Appl. Phys.,* vol. 91, pp. 595-604, 2002.
[http://dx.doi.org/10.1063/1.1425448]

[13] A. Mikami, and T. Koyanagi, "High efficiency 200-lm/W green light emitting organic devices prepared on high-index of refraction substrate", *SID 09 DIGEST,* pp. 907-910, 2009.

[14] Y-J. Lee, S-H. Kim, J. Huh, G-H. Kim, Y-H. Lee, S-H. Cho, Y-C. Kim, and Y.R. Do, "A high-extraction-efficiency nanopatterned organic light-emitting diode", *Appl. Phys. Lett.,* vol. 82, pp. 3779-3781, 2003.
[http://dx.doi.org/10.1063/1.1577823]

[15] K. Ishihara, M. Fujita, I. Matsubara, T. Asano, S. Noda, H. Ohata, A. Hirasawa, H. Nakada, and N.

Shimoji, "Organic light-emitting diodes with photonic crystals on glass substrate fabricated by nanoimprint lithography", *Appl. Phys. Lett.,* vol. 90, p. 111114, 2007.
[http://dx.doi.org/10.1063/1.2713237]

[16]   Y. Xiao, J.P. Yang, P.P. Cheng, J.J. Zhu, Z.Q. Xu, Y.H. Deng, S.T. Lee, Y.Q. Li, and J.X. Tang, "Surface plasmon-enhanced electroluminescence in organic light-emitting diodes incorporating Au nanoparticles", *Appl. Phys. Lett.,* vol. 100, p. 013308, 2012.
[http://dx.doi.org/10.1063/1.3675970]

[17]   X-B. Shi, C-H. Gao, D-Y. Zhou, M. Qian, Z-K. Wang, and L-S. Liao, "Surface plasmon polariton enhancement in blue organic light-emitting diode: role of metallic cathode", *Appl. Phys. Express,* vol. 5, p. 102102, 2012.
[http://dx.doi.org/10.1143/APEX.5.102102]

[18]   Y. Gu, D-D. Zhang, Q-D. Ou, Y-H. Deng, J-J. Zhu, L. Cheng, Z. Liu, S-T. Lee, Y-Q. Li, and J-X. Tang, "Light extraction enhancement in organic light-emitting diodes based on localized surface plasmon and light scattering double-effect", *J. Mater. Chem. C Mater. Opt. Electron. Devices,* vol. 1, pp. 4319-4326, 2013.
[http://dx.doi.org/10.1039/c3tc30197d]

[19]   H.J. Peng, C.F. Qiu, Z.L. Xie, H.Y. Chen, M. Wong, and H.S. Kwok, "Optical simulation of top-emitting organic light emitting diodes", *ASID,* vol. 8.3, no. 3-89, pp. 331-334, 2004.

[20]   B.W. D'Andrade, and J.J. Brown, "Organic light-emitting device luminaire for illumination applications", *Appl. Phys. Lett.,* vol. 88, no. 192908, 2006.

[21]   T. Yamasaki, K. Sumioka, and T. Tsutsui, "Organic light-emitting device with an ordered monolayer of silica microspheres as a scattering film", *Appl. Phys. Lett.,* vol. 76, pp. 1243-1245, 2000.
[http://dx.doi.org/10.1063/1.125997]

[22]   R. Bathelt, D. Buchhauser, C. Garditz, R. Paetzold, and P. Wellmann, "Light extraction from OLEDs for lighting applications through light scattering", *Org. Electron.,* vol. 8, pp. 293-299, 2007.
[http://dx.doi.org/10.1016/j.orgel.2006.11.003]

[23]   J.J. Shiang, T.J. Faircloth, and R. Anil, "Duggal, Experimental demonstration of increased organic light emitting device output *via* volumetric light scattering", *J. Appl. Phys.,* vol. 95, pp. 2889-2895, 2004.
[http://dx.doi.org/10.1063/1.1644038]

[24]   C-C. Liu, S-H. Liu, K-C. Tien, M-H. Hsu, H-W. Chang, C-K. Chang, C-J. Yang, and C-C. Wu, "Microcavity top-emitting organic light-emitting devices integrated with diffusers for simultaneous enhancement of efficiencies and viewing characteristics", *Appl. Phys. Lett.,* vol. 94, p. 103302, 2009.
[http://dx.doi.org/10.1063/1.3097354]

[25]   G. Nenna, A. De Girolamo DelMauro, E. Massera, A. Bruno, T. Fasolino, and C. Minarini, "Optical properties of polystyrene-ZnO nanocomposite scattering layer to improve light extraction in organic light-emitting diode", *J. Nanomater.,* vol. 2012, pp. 1-7, 2012.
[http://dx.doi.org/10.1155/2012/319398]

[26]   S. Chen, and H.S. Kwok, "Light extraction from organic light-emitting diodes for lighting applications by sand-blasting substrates", *Opt. Express,* vol. 18, no. 1, pp. 37-42, 2010.
[http://dx.doi.org/10.1364/OE.18.000037] [PMID: 20173819]

[27]   J-W. Shin, D-H. Cho, J. Moon, C.W. Joo, S.K. Park, J. Lee, J-H. Han, N.S. Cho, J. Hwang, J.W. Huh, H.Y. Chu, and J-I. Lee, "Random nano-structures as light extraction functionals for organic light-emitting diode applications", *Org. Electron.,* vol. 15, pp. 196-202, 2014.
        [http://dx.doi.org/10.1016/j.orgel.2013.11.007]

[28]   T. Bocksrocker, J. Hoffmann, C. Eschenbaum, A. Pargner, J. Preinfalk, F. Maier-Flaig, and U. Lemmer, "Micro-spherically textured organic light emitting diodes: a simple way towards highly increased light extraction", *Org. Electron.,* vol. 14, pp. 396-401, 2013.
        [http://dx.doi.org/10.1016/j.orgel.2012.10.036]

[29]   Y. Jiang, S. Chen, G. Li, H. Li, and H-S. Kwok, "A low-cost nano-modified substrate integrating both internal and external light extractors for enhancing light out-coupling in organic light-emitting diodes", *Adv. Optical Mater.,* vol. 2, pp. 418-422, 2014.
        [http://dx.doi.org/10.1002/adom.201400012]

[30]   C.S. Choi, S-M. Lee, M.S. Lim, K.C. Choi, D. Kim, D.Y. Jeon, Y. Yang, and O.O. Park, "Improved light extraction efficiency in organic light emitting diodes with a perforated WO3 hole injection layer fabricated by use of colloidal lithography", *Opt. Express,* vol. 20, suppl. Suppl. 2, pp. A309-A317, 2012.
        [http://dx.doi.org/10.1364/OE.20.00A309] [PMID: 22418680]

[31]   S.J. Shin, T.H. Park, J.H. Choi, E.H. Song, H. Kim, H.J. Lee, J-I. Lee, H.Y. Chu, K.B. Lee, Y.W. Park, and B-K. Ju, "Improvement of light out-coupling in organic light-emitting diodes by printed nanosized random texture layer", *Org. Electron.,* vol. 14, pp. 187-192, 2013.
        [http://dx.doi.org/10.1016/j.orgel.2012.10.009]

[32]   J.Y. Kim, W.H. Kim, D.H. Kim, and K.C. Choi, "Investigation of voltage reduction in nanostructure-embedded organic light-emitting diodes", *Org. Electron.,* vol. 15, pp. 260-265, 2014.
        [http://dx.doi.org/10.1016/j.orgel.2013.11.019]

[33]   M. Boroditsky, T.F. Krauss, R. Coccioli, R. Vrijen, R. Bhat, and E. Yablonovitch, "Light extraction from optically pumped light-emitting diode by thin-slab photonic crystals", *Appl. Phys. Lett.,* vol. 75, pp. 1036-1038, 1999.
        [http://dx.doi.org/10.1063/1.124588]

[34]   D. Stegall, S. Lamansky, J. Anim-Addo, M. Gardiner, E. Hao, L. Kreilich, F.B. McCormick, H. Le, Y. Lu, T.L. Smith, D. Wang, and J-Y. Zhang, "OLED light extraction with roll-to-roll nanostructured films", *Proc. SPIE,* vol. 7415, pp. 1-8, 2009.
        [http://dx.doi.org/10.1117/12.829396]

[35]   B.V. Enustun, and J. Turkevich, "Coagulation of colloidal gold", *J. Am. Chem. Soc.,* vol. 85, pp. 3317-3328, 1963.
        [http://dx.doi.org/10.1021/ja00904a001]

[36]   C. Wang, D. van der Vliet, K-C. Chang, H. You, D. Strmcnik, J.A. Schlueter, N.M. Markovic, and V.R. Stamenkovic, "Monodisperse Pt_3Co nanoparticles as a catalyst for the oxygen reduction reaction: size-dependent activity", *J. Phys. Chem. C.,* vol. 113, pp. 19365-19368, 2009.
        [http://dx.doi.org/10.1021/jp908203p]

CHAPTER 4

Encapsulation Technologies

Yibin Jiang[*], Shuming Chen

Department of Electronic and Computer Engineering, The Hong Kong University of Science and Technology, Hong Kong

Abstract: Flexible organic light emitting diode (OLED) display is an exciting and attractive technology to the consumers and the manufacturers, but there still exists a few challenges before it comes into real application. The biggest challenge is the demanding water and oxygen permeation requirement of the encapsulation, since the lifetime of OLED will drastically decrease when it is exposed to moisture and oxygen. As the barrier performance of polymer is not as good as that of glass, both sides of the OLED device which is deposited on polymer substrate need to be encapsulated by dense thin films. This chapter will provide a summary of the encapsulation technologies, including the traditional encapsulation and the advanced thin film encapsulation. First, the degradation mechanism, the permeation mechanism, and the permeation measurement are introduced as background information. Then thin film encapsulation technologies (Vitex organic/inorganic multilayer and atomic layer deposited film) will be the main topic.

Keywords: Atomic layer deposition, Barrier, Ca test, Dark spots, Degradation, Encapsulation, Flexible display, Getter, Glass or metal lid, Lifetime, Moisture, Multilayer, Nanocomposite film, Oxygen, Permeation, Water vapor transmission rate.

INTRODUCTION

The organic materials, which make up different functional layers of OLEDs, are very sensitive to the oxygen and moisture. Without encapsulation, non-emitting areas which are called "dark spots" will quickly appear once the OLEDs are exposed to atmosphere environment. The dark spots will gradually grow larger and larger, and finally cover the entire device area of the OLEDs and make the device "die" several tens of hours later. Therefore, encapsulation technologies are

[*] **Corresponding Author Yibin Jiang:** Department of Electronic and Computer Engineering, The Hong Kong University of Science and Technology, Hong Kong; Tel.: +852-23588845; E-mail: yjiangaa@connect.ust.hk

very necessary in order to increase the lifetime of the OLED device, considering the long lifetime requirements in the OLED display and OLED lighting applications.

The existing encapsulation technologies can be classified into two categories: traditional encapsulation and thin film encapsulation. This chapter is divided into three sections. The encapsulation requirements will be discussed firstly. Then the traditional encapsulation technology will be covered. Finally, the state-of-art thin film encapsulation technology will be the theme of the last section.

DARK SPOTS FORMATION MECHANISM

Dark spots formation is a story about the defects in OLED device. Therefore defects should be talked about first. It is well known that the organic material forms continuous film without pinholes, while the inorganic material grows film with pinholes due to the granular structure of the material, the dust particle contamination on the substrate or the protrusions of the substrate. As a result, pinhole defects are formed on top of the organic layers after the metal deposition in OLED [1], as shown in Fig. (**4.1**). It has been found that the distribution of the dark spot area follows a Gauss distribution just as the typical dust size distribution [2]. This indicates that dust contamination may be the major defects sources. Apart from the pinhole defects, there are inherent natural defects which are the device edges defined by the metal electrode due to the device structure.

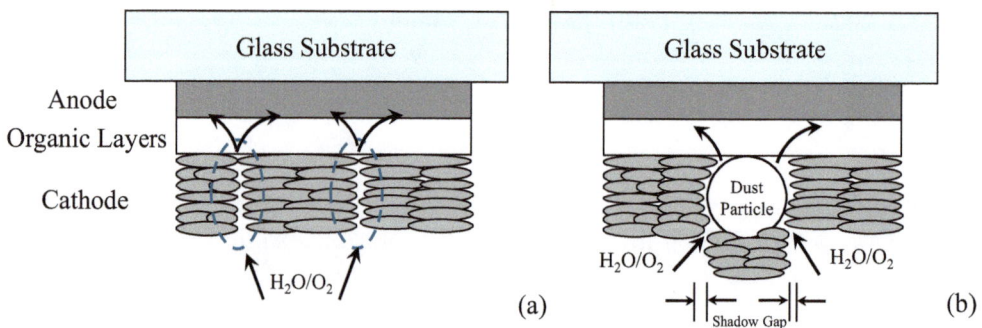

Fig. (4.1). Proposed mechanism of the formation of pinhole defects (**a**) at grain boundaries and (**b**) at contaminant particles.

Water vapor and oxygen can diffused to the organic layers through these defects as shown in Fig. (**4.2**). After the water vapor and oxygen flow down to the bottom of the hole, they begin to diffuse laterally into the interface between the cathode and the organic layers through the boundary of the hole, as indicated by the white arrow. Therefore, the hole perimeter determines the amount of water and oxygen diffusing into the interface at a certain time. As a result, the area and growth rate of the dark spots show a linear relationship with the pinhole diameter.

Under the normal operation of OLEDs, the water on the cathode-organic interface will undergo an electrochemical reduction process because of the electrode field across the organic layers. This electrochemical process leads to hydrogen gas evolution according to: $2H_2O + 2e^- \rightarrow H_2 + 2OH^-$.

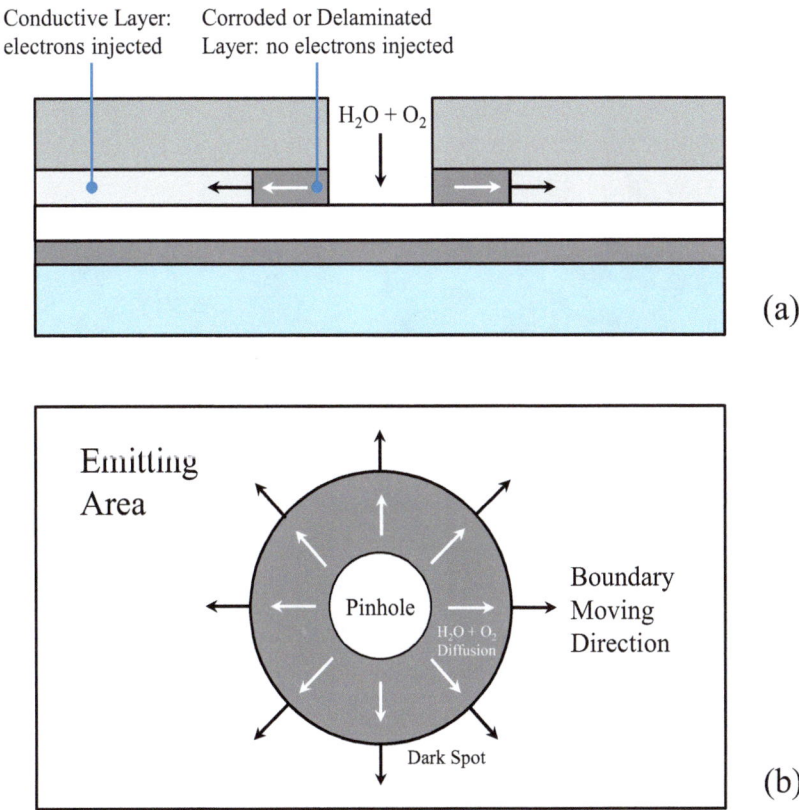

Fig. (4.2). Schematic drawings showing the process and mechanism of (**a**) dark spot formation and (**b**) growth.

Under continuous operation, the amount of hydrogen will increase, and then bubbles are created due to gas pressure, lifting off the cathode. Fig. (**4.3a**) shows the schematic Because of this cathode delamination, electron cannot be injected into the organic layers from the cathode, thus leaving non-emitting dark spot underneath. The dark spot growth process is schematic shown in Fig. (**4.3a**). Water vapor can not only initiate dark spots, but also accelerate the growth rate of dark spots. This is because that additional entrance of water vapor will be induced when the bubbles keep growing and finally burst. Besides, thin LiF, which is widely used between the organic layers and the cathode as an electron injection layer, can also enhance the growth rate in two ways. First, LiF will easily absorb water and then promote water's diffusion under the cathode because of its hydrophilic nature. Secondly, the salt will be solvated by water and form electrolyte which promote the electrochemical reduction process of water.

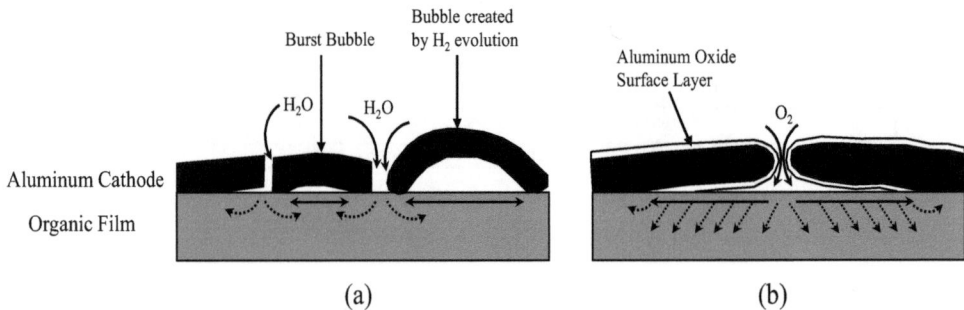

Fig. (**4.3**). (**a**) Dark spot growth process of a driving OLED under a water vapor atmosphere. (**b**) Dark spot growth process of an unbiased OLED under pure a pure oxygen atmosphere.

It has been observed that there still exists dark spot growth on the device even in pure oxygen atmosphere [3]. This means that oxygen, which also diffuses through the defects, is another degradation factor of the OLED device. It has been proved that oxidation of the cathode results in dark spots appearance and growth, while oxidation of the organic material leads to device efficiency decrease [4]. Furthermore, photo-oxidation reaction will result in organic material decomposition, which may produce volatile species. When the oxygen diffuses to the cathode-organic interface, a 2/nm thick surface oxide layer will be gradually grown in less than an hour at room temperature [5]. When the metal is oxidized to be oxide, the volume of the material will expand to be around 1.5 times larger during the process. Thus the chemical conversion and the volumetric expansion will give rise to cathode delamination, inhibiting charge injection as schematic shown in Fig. (**4.3b**).

REQUIREMENT AND MEASUREMENT OF THE PERMEATION RATES

The exact permeation requirement of the encapsulation for OLED is unclear, as the long-term degradation mechanism is still under debate. However, an acceptable upper limit of the moisture permeation rate can be calculated by simply assuming that oxidation of the cathode is the major lifetime-limiting factor. For example, assuming an Al cathode with thickness of 700 Å, density 2.70 g/cm^3 and molar mass 27 g/mol, then an OLED device contains 7×10^{-7} mol/cm^2 Al. About 1.9×10^{-5} g H_2O can completely oxidize one square centimeter such cathode. Thus, an average moisture permeation rate of 5.2×10^{-4} $g/m^2/day$ is required to achieve a one year device lifetime. In reality, the cathode-organic interface rather than the bulk cathode determines the device lifetime. Therefore, 10 % oxidation of the cathode can cause severe device degradation. That means that the above permeation rate should be reduced one order of magnitude to 5.2×10^{-5} $g/m^2/day$. Furthermore, taking into account the potential organic material degradation by water, an upper limit of 1×10^{-6} $g/m^2/day$ can be estimated as the moisture permeation rate requirement for the encap-sulation.

The 1×10^{-6} $g/m^2/day$ water vapor transmission rate (WVTR) requirement is extremely low compared to other application requirements as shown in Fig. (**4.4**) [6]. Actually, this rate is below the low limit of the standard moisture permeation measurement device from Brugger [7]. This device can measure WVTR above 0.005 $g/m^2/day$. The mechanism of detecting the permeated water vapor is electrolysis. Therefore, new WVTR measurement method, which can preciously measure as low as 1×10^{-6} $g/m^2/day$, should be developed in order to evaluate and characterize the encapsulation technologies for OLED.

One method is called the "Ca test", as illustrated in Fig. (**4.5a**). Calcium is a reactive metal. When the opaque Ca reacts with water or oxygen, it will convert to a transparent insulator. This optical and electrical change can be utilized to measure the transmission of O_2 and H_2O. The resistance change of the Ca under a constant voltage allows the determination of the O specie transmission rate with high degree accuracy. The reported WVTR using Ca test is as low as 3×10^{-7} $g/m^2/day$ [8]. Although the Ca test has the high accuracy feature, it can't distinguish between oxygen and water permeation. This means that the measured rate is the maximum value for WVTR or OTR.

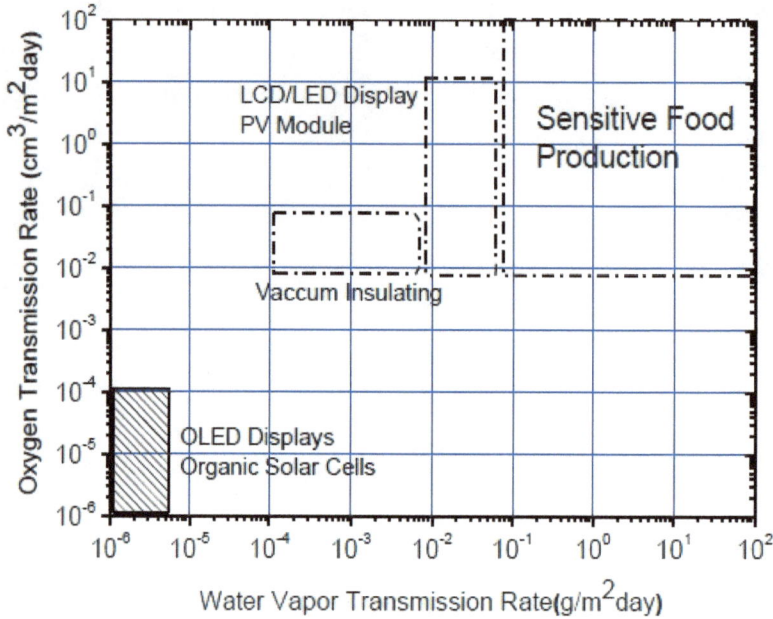

Fig. (4.4). The OTR and WVTR requirements for the encapsulation of different products [6].

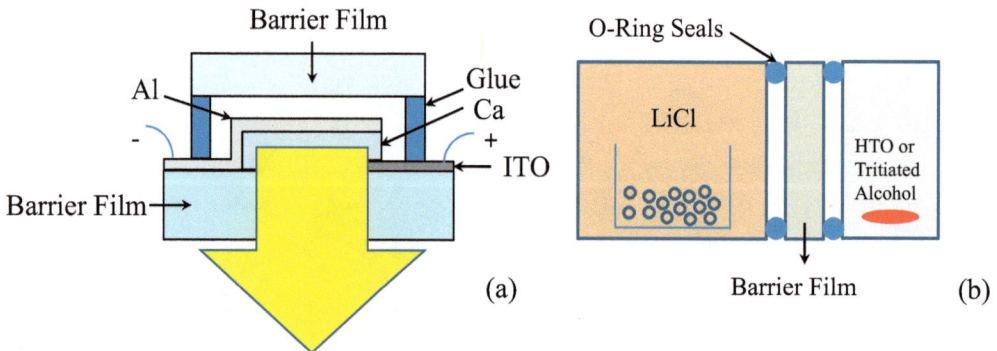

Fig. (4.5). The permeation measurement setup: (**a**) calcium test, (**b**) tritiated water test.

Another famous approach to reliably measure water permeation is the tritium trans-mission rate (TTR) measurement, which uses the radioactive HTO (hydrogen-tritium-oxygen) tracer [9 - 11]. Fig. (**4.5b**) describes the TTR measurement: after HTO, which con-tain the tritium atom, permeates through the barrier film, the LiCl will absorb the HTO molecule; the decays of the absorbed tritium atoms can be counted by a scintillation counter and then used to calculate the HTO permeation rate. The WVTR measurement limit of this method can be as low as 1×10^{-6} g/m^2/day.

TRADITIONAL ENCAPSULATION TECHNOLOGY

Most work of the early day's OLED R&D was based on glass substrate. As a result, the encapsulation can be realized by sealing the device under a glass or metal lid using UV-cured epoxy resin, as shown in Fig. (**4.6a**).

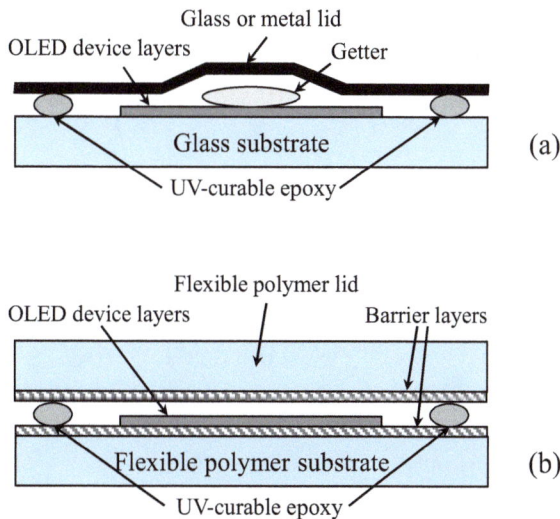

Fig. (4.6). Structures of the traditional OLED encapsulation using: (**a**) solid lid together with getter, or (**b**) flexible barrier-coated polymer lid.

A getter like calcium oxide or barium oxide needs to be added before the curing in order to absorb byproducts which are produced during the resin cure process or water that diffuses through the epoxy seal over time. The entire encapsulation process needs to be completed in inert atmosphere such as argon or nitrogen. The traditional encapsulation is relatively simple and low cost. But this technology can't be applied to flexible OLED displays due to the rigidity of the lid. In order to apply this encapsulation technology to flexible OLED (FOLED) displays, people made a little change by replacing the rigid lids with barrier-coated flexible lids as shown in Fig. (**4.6b**). No getter needs to be added. But there exists a gap between the polymer substrate and the lid for this design, thus the lid may have the potential to cause abrasion damage during the bending or folded use. Besides, typically epoxy sealing is rigid, so flexible sealing needs to be developed.

THIN FILM ENCAPSUTION TECHNOLOGY

This is another encapsulation technology for FOLED displays. As illustrated in Fig. (**4.7**), thin film barrier coating intimately covers the OLED device, rendering this technology light, thin and abrasion-free features. However, this technology requires that the thin-film barrier deposition process should not cause any damage to the active materials of the OLED beneath, *i.e.*, the process temperature should be low enough to be compatible with the organic materials and any exposure to solvents must be avoided. Therefore, TFE can't give the manufacturer a great range of choices for the deposition technology because of limited process conditions compared to the barrier coated flexible lid encapsulation.

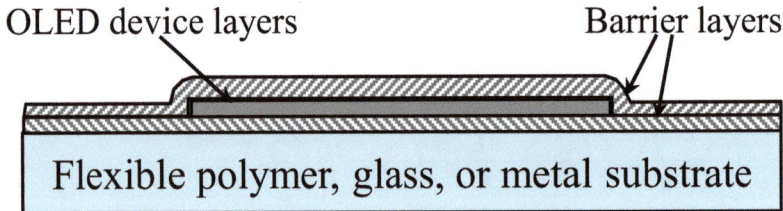

Fig. (4.7). Schematic diagrams of monolithic thin film encapsulation for OLED.

Before talking about different TFE technologies, the permeation properties of different thin films should be reviewed firstly. Thin film can be categorized into two kinds: inorganic and organic thin films. Organic materials usually form films with low density. Due to this low dense feature, the WVTRs of most organic film are ~10 g/m2/day. For inorganic material, the moisture & oxygen permeation rates of its thin film form are orders of magnitude higher comparing to its dense bulk counterpart. The reason is that it is the defects of the film rather than the material itself that dominate the vapor per-meation. Table **4.1** gives a summary of the defect properties of different inorganic films deposited on polymeric substrates [12]. It can be observed from this table that aluminum films grown by different deposition methods have different defect properties, indicating the importance of the fabrication method. Plasma-enhanced chemical vapor deposition (PECVD) can grow conformal coverage films as exhibited in Fig. (**4.8a**) due to the vapor phase nature of its source, while physical vapor deposition (PVD) such as evaporation or sputtering can only form nonconformal coverage films as shown in Fig. (**4.8b**) because of the shadow effect. The shadow effect is illustrated in Fig. (**4.9**). In PVD process, a point source is used and atoms of the material are radially evaporated or sputtered from this source. The atoms go in straight line, travel the distance between the source and the substrate, finally deposit onto the substrate. If there are any features such as spikes or pores on the substrate, the features will shadow the atoms from reaching the substrate and cause the so-called shadow effect. Besides, the films grown by PVD show larger and denser defects than those by PECVD because the deposition process is usually accompanied with arc discharging and energetic particles. The best single-layer barrier performance for inorganic film is achieved by PECVD.

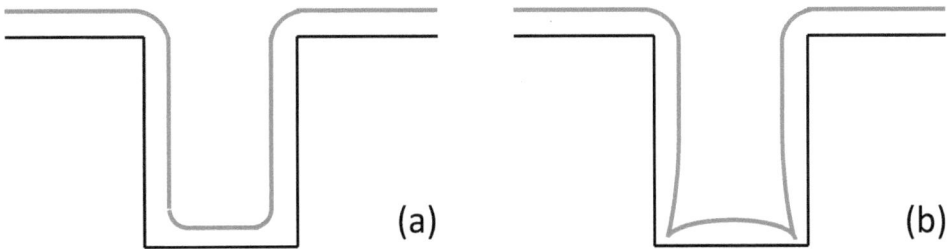

Fig. (**4.8**). (**a**) conformal and (**b**) non-conformal coverage.

Table 4.1. Summary of the average defect size and defect density for single-layer inorganic films deposited on polymeric substrates with different technologies.

Material	Defect radius (µm)	Defect density (mm^{-2})	Deposition method	Substrate
SiO$_2$	0.6	11-1100	PECVD	PET
Si$_3$N$_4$	0.6	5-1000	PECVD	PET
Al	1	25-400	Evaporation	PET
Al	2-3	200	Sputtering	PET
AlO$_x$	1	700	Sputtering	PET

(Table 4.1) contd.....

Material	Defect radius (µm)	Defect density (mm^{-2})	Deposition method	Substrate
AlO$_x$N$_y$	0.5-1.4	600	Sputtering	PET

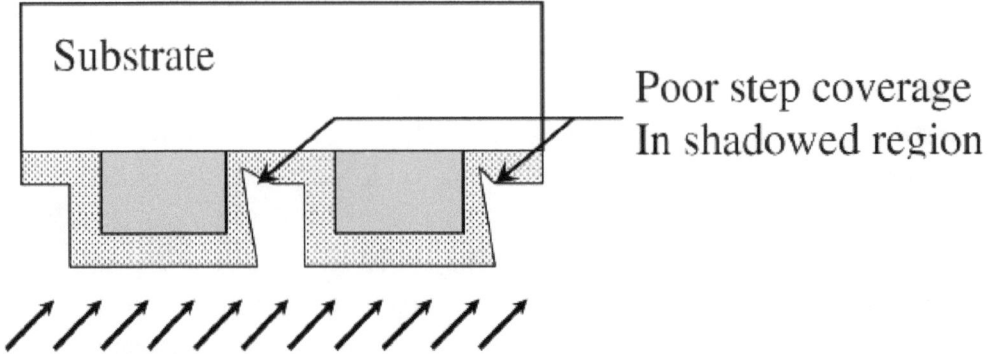

Fig. (4.9). Illustration of shadow effect due to collimated flux.

Fig. (**4.10a**) and (**4.10b**) shows the OTR and WVTR of SiO$_2$ thin film *versus* coating thickness. OTR changes little with increasing film thickness until a certain critical dc thickness is reached. At dc the OTR value drops by about two orders of magnitude. For d>dc, OTR still decreases, but at a slow rate. The behavior of WVTR is a little different from that of OTR: the WVTR decreases quickly before dc , then it decreases slowly. The behaviors of OTR and WVTR are symptomatic of defects dominated permeation. Table **4.2** summarizes the WVTRs of different inorganic thin films coated on polymeric substrate.

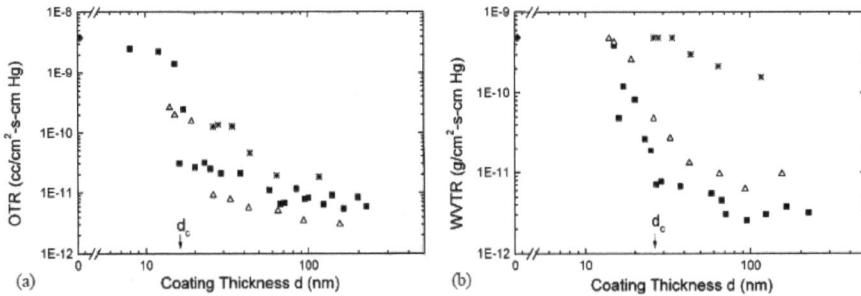

Fig. (4.10). (**a**) Oxygen permeation rate (OTR) and (**b**) water vapor permeation rate (WVTR) as a function of coating thickness, d, for SiO$_2$ deposited on 13µm PET.

Table 4.2. Permeation of water through polymeric substrate coated with different thin films.

Material	Film thickness (nm)	WVTR (g/m²/day)
SiO_2	35	0.5
Si_3N_4	35	0.2
Ta_2O_5	50	0.7
ITO	50	0.4
Al	50	0.1
AlO_x	35	0.4

According to the above table, although various high density materials, including silicon nitride, silicon oxide, aluminum oxide and tantalum pentoxide, have been investigated, a single-layer barrier film still shows in orders of magnitude short of the WVTR requirement of OLEDs. However, the barrier performance can be significantly enhanced just with periodic multilayer structure of two materials. Up to now, a lot of multilayer structures including inorganic/inorganic multi-stacks and organic/inorganic multi-stacks have been proposed.

Si_3N_4/SiO_2 Multilayer

The high permeability of single-layer inorganic film was originated from the pinhole defects. From Table **4.1**, it can be observed that the pinhole defect density of PECVD deposited SiO_2 and Si_3N_4 varies in a big range. This is caused by the difference of the deposition environment temperature. The traditional process temperature of PECVD is 300 °C, under which the deposited film shows few pinhole defects. However, this temperature is higher than the glass transition temperature (T_g) of organic materials and will cause the morphological change of the OLED's active layers over time and finally damage the device. Therefore, the process temperature of PECVD should be reduced in order to be compatible with OLEDs. But the pinhole density will increase at low deposition temperature. In order to solve this contradiction, Philips proposed to stack multi Si_3N_4 and SiO_2 which are deposited at 85 °C [13]. SiO_2 can changes the chemical interface of the below defective Si_3N_4 film and decouple its pinhole defects from influencing the growth of the above Si_3N_4 film. As a result, the pinhole defects can't continuously grow and the number of pinholes can be reduced. Stacking more layers can reduce the number of pinholes even further, and with proper process condition, even zero pinhole can be possible. The pinhole test of this stack together with other reference films is shown in Table **4.3** [13]. The pinhole test was conducted by inserting barrier coated aluminum wafer into the Kings water (3 parts 37% HCl, 1 part 65% HNO_3). This solution etches Al, but not Si_3N_4 or SiO_2. As the viscosity of the etchant solution is very low, the etchant can penetrate through the pinholes easily. After etching, large holes can then be indicated by the place where Al was etched. The Ca test was also conducted to further confirm the Si_3N_4/SiO_2 multilayer's high barrier performance. It was found that the permeability of a single $Si_3N_4/SiO_2/Si_3N_4$ stack deposited at 85 °C is less than 1×10^{-6} g/m²/day [13]. Fig. (**4.11**) shows the steps of the Ca test process [13].

Table 4.3. Pinhole test result of different films deposited at different temperatures.

PECVD layer	Deposition temperature (°C)	Pinholes
SiO_2	300	few
SiON	300	no
Si_3N_4	300	no
SiO_2	85	some
SiON	85	many
Si_3N_4	85	few
$Si_3N_4/SiO_2/Si_3N_4$	85	no

Fig. (4.11). Photos of a calcium test during different stages of the permeation process.

The $Si_3N_4/SiO_2/Si_3N_4$ stack has a good step coverage which is attributed to the conformal coverage property of PECVD. This good step coverage makes it possible to completely cover and seal particles. The step coverage of the $Si_3N_4/SiO_2/Si_3N_4$ stack is exhibited in Fig. (**4.12**). It can be found that the thickness of the deposited film on the negative angle slope is about 50% of that of the film on top of the step.

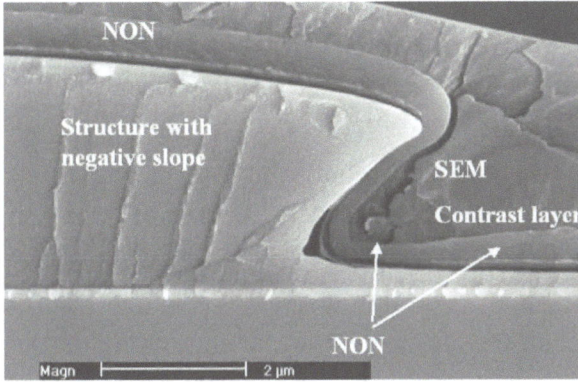

Fig. (4.12). Step coverage of a $Si_3N_4/SiO_2/Si_3N_4$ stack over a negative structure.

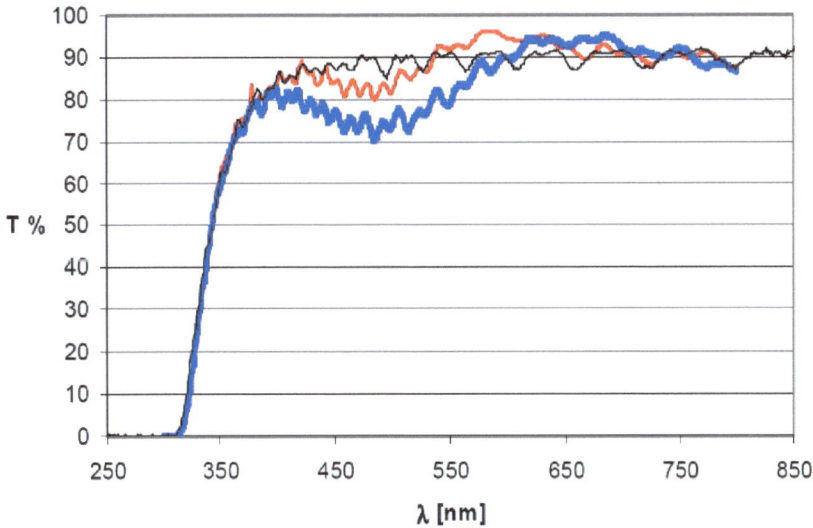

Fig. (4.13). Comparison of the transmission spectra of a PES plastic substrate (black, thin line), a $Si_3N_4/SiO_2/Si_3N_4$ stack on PES (red) and $Si_3N_4/SiO_2/Si_3N_4/SiO_2/Si_3N_4$ stack on PES (blue, thick line).

The transparency of the thin film barrier is another characteristic need to be considered when applying to top emission devices or transparent devices. As shown in Fig. (**4.13**), the transparency of the $Si_3N_4/SiO_2/Si_3N_4$ stack is greater than 80% over the entire visible range. Adding one more pair, the transparency of a $Si_3N_4/SiO_2/Si_3N_4/SiO_2/Si_3N_4$ stack on PES decreases slightly, but still above 70%.

Organic/Inorganic Multilayer

Nowadays, alternating organic/inorganic multilayer is suggested as another TFE solution for OLEDs. Vitex System [14 - 16] is one of the pioneers in this area and they reported encapsulation (trade name of Barix) made of alternating Al_2O_3 and polyacrylate layers as shown in Fig. (**4.14**) [12]. The Al_2O_3 layers are deposited onto the substrate by reactively sputtering an aluminum target, while the polyacrylate layers are formed *via* flash evaporation of the monomer followed by ultraviolet (UV) curing [12]. Since the polyacrylate deposition technique involves the deposition and curing of the liquid monomer which leads to non-conformal polymer film, this technique has the feature of smoothing the substrate. Thus, the polyacrylate layer can decouple any non-uniformity on the substrate, which is illustrated by the AFM images shown in Fig. (**4.15**) of a PET substrate before and after polyacrylate coating. Besides, the polyacrylate layer can also decouple the defects in the Al_2O_3 layer. That is to say, the defects in the underlying Al_2O_3 layer can't propagate to the upper layer. Thereby this multilayer structure can lower the pinhole density.

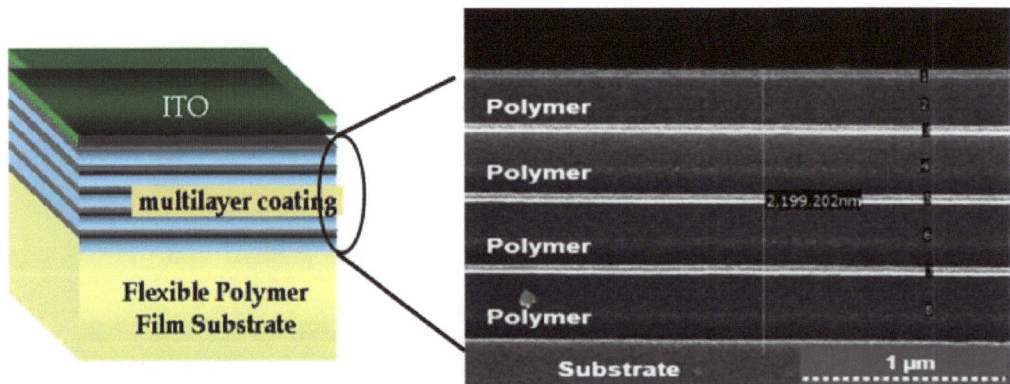

Fig. (4.14). A cross-sectional scanning electron microscopy (SEM) of the multilayer barrier structure (inorganic/organic layers): four pairs of AlO_x/polymer multilayer deposited by the Vitex barrier deposition system.

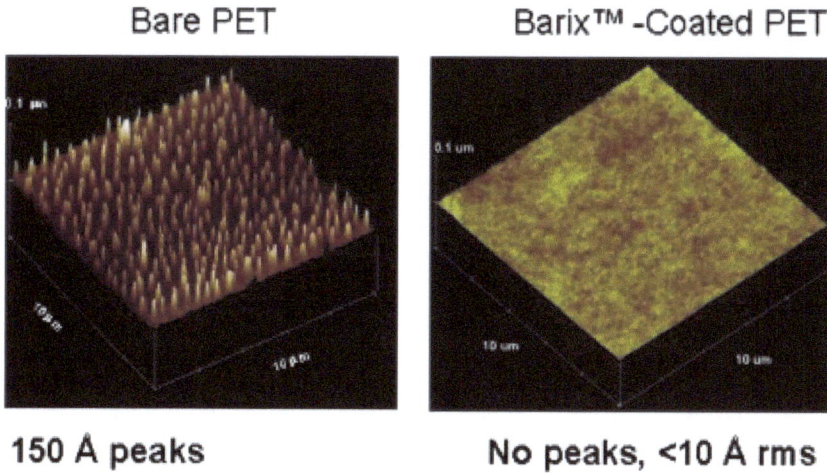

Fig. (4.15). AFM of (**a**) a commercially available polyester substrate, and (**b**) an identical substrate with a 0.25μm thick polyacrylate layer deposited *via* roll-coating in vacuum.

When the polymer layer thickness is larger than the average size of pinholes of the inorganic layer (Al_2O_3), the gas will permeate through a tortuous path which can provide a longer diffusion path. Therefore, the WVTR of this multilayer structure has a strong dependence on the organic layer thickness. With the lowered pinhole density and lengthened diffusion path, the barrier performance of this multilayer structure is greatly improved. With Ca test, the WVTR of this polyacrylate/Al_2O_3 multilayer approaches the 10^{-6} g/m²/day. As can be seen in Fig. (**4.16**), water permeation can be dramatically red-uced with three pairs of polyacrylate/Al_2O_3 layer stacks [16].

Other materials were also investigated for this organic/inorganic multilayer scheme as shown in Table (**4.4**). The barrier performances of these listed multilayer thin films are quite close to the permeation requirement of the OLEDs. In the organic/inorganic multilayer TFE, organic layers are used to smooth the surface, decouple the pinhole defects and increase the diffusion length. However, most organic layer deposition techniques involve vacuum-based process which requires considerable time and increases the cost. Therefore, Bae's group synthesized an epoxy hybrid material (hybrimer) called cycloaliphatic-epoxy oligosiloxane resin which can be simply coated with a solution process [17]. The WVTR value of this hybrimer depends on the siloxane and epoxy network density, and the surface energy of the coating [18]. A WVTR value of 3.9 g/m²/day for the 4 μm was obtained. In order to further improve the barrier property, the same group incorporated silica nanoparticles into the polymer matrix [19]. The improvement of the gas barrier performance is mainly attributed to an extended diffusive path, since the penetrated gas must detour around the nanoparticles within the matrix [20]. The synthesis of this so-called silica nanoparticle-embedded hybrimer nano-composite (S-H nanocomposite) is shown in Fig. (**4.17**). As shown in Table **4.5**, the barrier performance of the S-H nanocomposite film coated on PET exhibits significant improvement with an increase in silica content (that is, WVTR from 3.56 g/m²/day for 10 wt% silica to 0.24 g/m²/day for 100 wt% silica).

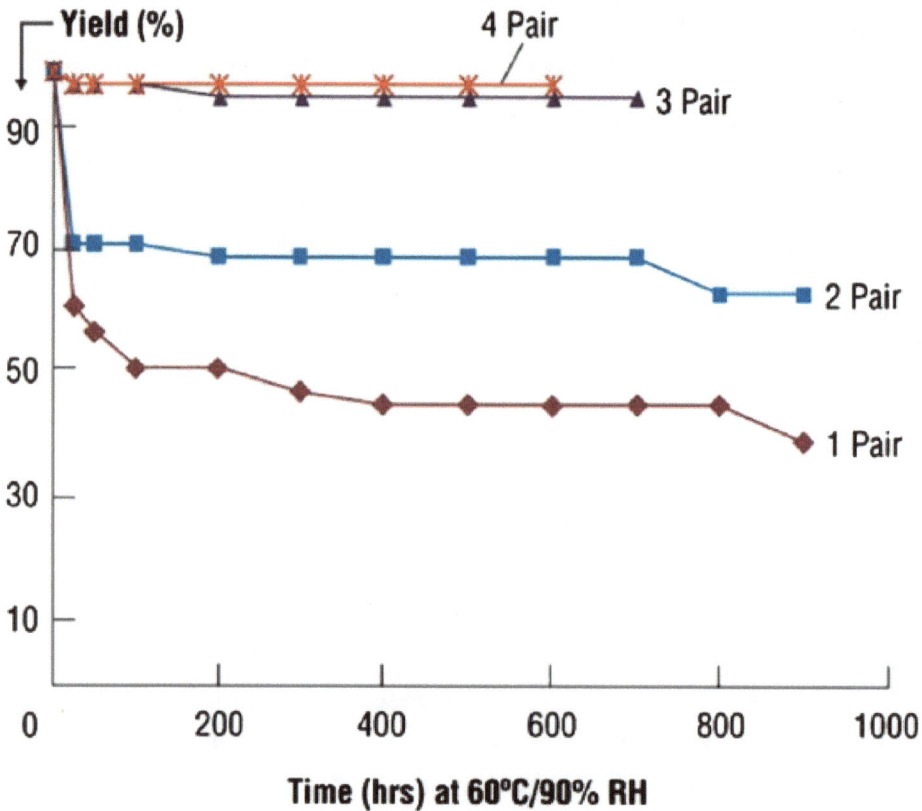

Fig. (4.16). Calcium test results for different number of inorganic/organic layers.

Table 4.4. Summary of the organic/inorganic combinations which have been investigated for the multilayer scheme encapsulation.

Structure	Organic layer	Inorganic layer	Number of stacks	WVTR (g/m²/day)	Transparency
Hybrimer/MgO	Polymerized cycloaliphatic epoxy hybrid material (hybrimer)	MgO	6	4.9×10^{-5}	81.8%
S-H nanocomposite/ MgO	Silica nanoparticle-embedded hybrid nanocomposite (S-H nanocomposite)	MgO	4	4.33×10^{-6}	84%
S-H nanocomposite/ Al₂O₃	S-H nanocomposite	Al₂O₃	3	1.14×10^{-5}	85.8%
Polyurea/Al₂O₃	Polyurea	Al₂O₃	5	5×10^{-4}	86%
CFₓ/Si₃N₄	Fluorocarbon (CFₓ)	Si₃N₄	5		
CFₓ/Al₂O₃:SiO₂	CFₓ	Al₂O₃:SiO₂	5	3.1×10^{-6}	70%

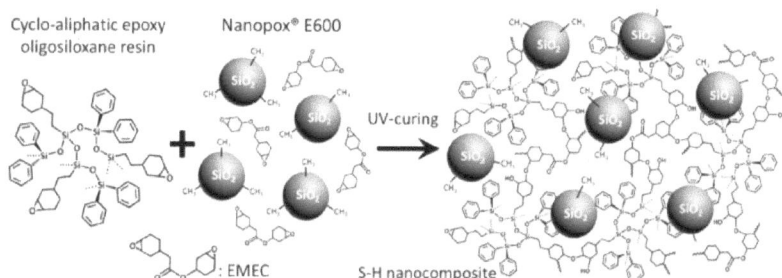

Fig. (4.17). S–H nanocomposite synthesized with a sol–gel cycloaliphatic oligosiloxane resin and Nanopox E600.

Table 4.5. Component formulation for different S-H nanocomposites and the corresponding WVTR values of S-H nanocomposites barrier coatings on PET film.

	Silica content (wt%)	Oligosiloxane resin (g)	Nanopox E600 (g)	Thickness (µm)	WVTR (g/m²/day)
Hybrimer	0	10	0	4	3.90
S-H nanocomposite	10	10	2.5		3.56
25	10	6.25			1.80
50	10	12.5			0.90
100	10	25			0.24

(Table 4.5) contd.....

	Silica content (wt%)	Oligosiloxane resin (g)	Nanopox E600 (g)	Thickness (μm)	WVTR (g/m²/day)
PET				100	24.4

Atomic Layer Deposited (ALD) Film

Atomic layer deposition is a conformal thin film deposition technique based on sequential, self-limiting surface reactions [12]. Fig. (**4.18**) shows the schematic diagram of the sequential, self-limiting surface reactions during the ALD process. Most ALD procedures make use of binary reactions where two sequential surface reactions are repeatedly performed to deposit a compound film. Chemically, ALD is just alike the chemical vapor deposition (CVD), except that ALD breaks down the whole reaction of CVD into the two sequential reactions of ALD — that is keeping the precursor materials separate.

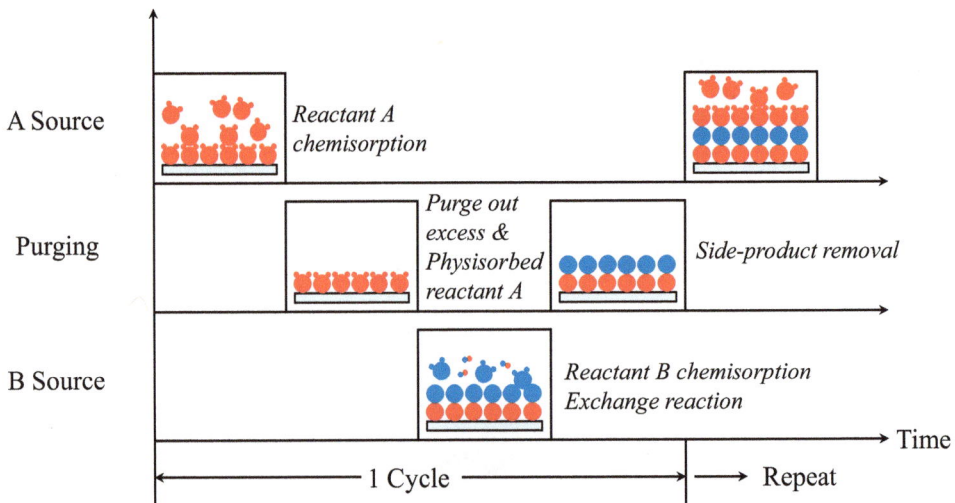

Fig. (4.18). Schematic representation of ALD using self-limiting surface chemistry and an AB binary reaction sequence.

Separation of the precursors is realized by injecting an inert gas (typically nitrogen or argon) after each reactant cycle to purge the residual precursor out of the process chamber, which is called "purging" process. Because the two surface reactions are self-limiting, then the two reactions may proceed in a sequential manner to deposit a thin conformal film with control of the thickness and composition at the atomic level [20, 21]. Besides, the self-limiting feature enables ALD to grow smooth, continuous and pinhole-free conformal film because the reactant gas can penetrate into every corner during the film growth. This advantage is very important for the TFE technology to enhance diffusion barrier property.

Summary of the barrier properties of the ALD thin films is given in Table (**4.6**). The ALD technology is very attractive to the manufacturers as its high quality film can meet the critical WVTR&OTR requirements from the highly sensitive OLED materials. Despite the great diffusion barrier performance, only one monolayer can be grown in one cycle. Thus, ALD's slow deposition rate drawback will limit the throughput when applied in large area fabrication. Recently, a few groups have developed an ALD system based on a new coating head design, which can work under the atmospheric pressure as shown in Fig. (**4.19**) [12]. The new head has many channels which are physically separated by inert gas flow. And the two ALD reactants are transported into the channels simultaneously. The head move back and forth in a certain speed to let two precursors react point by point and cycle by cycle. Because no purge and degas processes are involved in this ALD system, the film deposition rate can be improved. Therefore, this novel ALD system will provide a solution to overcome the limitation of high cost and low throughput on large area substrates, which make it possible for the future large-scale production of flexible display and lighting panels [22]

Table 4.6. Summary of the deposition conditions, structures and barrier properties of the ALD thin films [6].

Process	Material	Deposition condition	Thickness	WVTR (g/m²/day)	OLED Lifetime	Ref.
PEALD	Al_2O_3:N	TMA, O_2, N_2, temp:80 °C	300 nm	N/A	650h	[23]
ALD	Al_2O_3	TMA, H_2O, temp:120 °C	25 nm	1.7×10^{-5}	N/A	[24]
ALD	Al_2O_3	TMA, H_2O, temp:80 °C	30 nm	0.0615	193h	[25]

(Table 4.6) contd.....

Process	Material	Deposition condition	Thickness	WVTR (g/m²/day)	OLED Lifetime	Ref.
PEALD	Al_2O_3	TMA, O_2, temp:100 °C	10~40 nm	5×10^{-3}	N/A	[26]
PEALD	TiO_2	TDMAT, O_2, temp:90 °C	80	0.024	90h	[27]
ALD	Al_2O_3/SiO_x	TMA, Silanol, H_2O, temp:175 °C	5nm (Al_2O_3)/ 60nm (SiO_x)	5×10^{-5}	N/A	[28]
ALD	Al_2O_3/ZrO_2	TMA, TDMAZr, H_2O, temp:80 °C	2.1nm (Al_2O_3)/ 3.1nm (ZrO_2) total 100nm	4.75×10^{-5}	10000h	[29, 30]

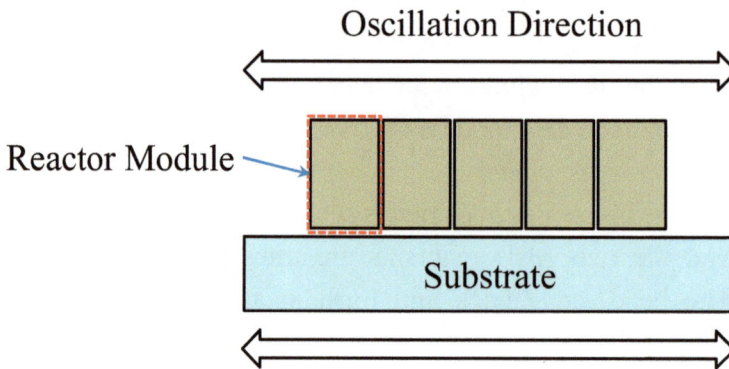

Fig. (4.19). A schematic diagram of the atmospheric pressure ALD system.

CONCLUSION

The markets and customers have waited a long time for the flexible OLED technology. But before this flexible diplay comes from the fiction into the daily life, further development of TFE technology is still required. Single barrier-layer structures are not sufficient for OLEDs, but the barrier performance can be greatly improved by adopting multilayer structures. The synergistic effect of organic and inorganic multilayer structures can lengthen the diffusion path, thus improving the barrier property. A lot of organic and inorganic materials have been investigated for this multilayer structure, and the barrier performance of this multilayer encapsulation has reached the 1×10^{-6} g/m²/day require-ment. However, the fabrication of this multilayer structure is complex and time-consuming. Recently, the ALD technology shows the promise to solve these issues of TFE. This chapter presents the encapsulation problems encountered in OLED fabrication and shows the corresponding solutions. Hope it be helpful along the progressing road of large area and flexible OLED technologies.

CONFLICT OF INTEREST

The authors confirm that this chapter contents have no conflict of interest.

ACKNOWLEDGEMENTS

Declared none.

REFERENCES

[1] J-S. Kim, P.K. Ho, C.E. Murphy, N. Baynes, and R.H. Friend, "Nature of Non-emissive black spots in polymer light-emitting diodes by *in-situ* micro-raman spectroscopy", *Adv. Mater.,* vol. 14, pp. 206-209, 2002.
[http://dx.doi.org/10.1002/1521-4095(20020205)14:3<206::AID-ADMA206>3.0.CO;2-J]

[2] S.F. Lim, W. Wang, and S.J. Chua, "Understanding dark spot formation and growth in organic light-emitting devices by controlling pinhole size and shape", *Adv. Funct. Mater.,* vol. 12, pp. 513-518, 2002.
[http://dx.doi.org/10.1002/1616-3028(20020805)12:8<513::AID-ADFM513>3.0.CO;2-7]

[3] M. Schaer, F. Nüesch, D. Berner, W. Leo, and L. Zuppiroli, "Water vapor and oxygen degradation mechanisms in organic light emitting diodes", *Adv. Funct. Mater.,* vol. 11, pp. 116-121, 2001.
[http://dx.doi.org/10.1002/1616-3028(200104)11:2<116::AID-ADFM116>3.0.CO;2-B]

[4] L. Ke, S-J. Chua, K. Zhang, and N. Yakovlev, "Degradation and failure of organic light-emitting devices", *Appl. Phys. Lett.,* vol. 80, pp. 2195-2197, 2002.
[http://dx.doi.org/10.1063/1.1464216]

[5] O. Kubaschewski, and B.E. Hopkins, *Oxidation of metals and alloys.* Butterworths: London, 1962, p. 38.

[6] J-S. Park, H. Chae, H.K. Chung, and S.I. Lee, "Thin film encapsulation for flexible AMOLED: a review", *Semicond. Sci. Technol.,* vol. 26, p. 034001, 2011.
[http://dx.doi.org/10.1088/0268-1242/26/3/034001]

[7] C. Charton, N. Schiller, M. Fahland, A. Hollander, A. Wedel, and K. Noller, "Development of high barrier films on flexible polymer substrates", *Thin Solid Films,* vol. 502, pp. 99-103, 2006.
[http://dx.doi.org/10.1016/j.tsf.2005.07.253]

[8] G. Nisato, P. C. P. Bouten, P. J. Slikkerveer, W. D. Bennett, G. L. Graff, N. Rutherford, and L. Wiese, *Proc. Asia Display/IDW,* p. 1435, 2001.

[9] M.D. Groner, S.M. George, R.S. McLean, and P.F. Carcia, "Gas diffusion barriers on polymers using atomic layer deposition", *Appl. Phys. Lett.,* vol. 88, p. 051907, 2006.
[http://dx.doi.org/10.1063/1.2168489]

[10] A.R. Coulter, R.A. Deeken, and G.M. Zentner, "Water permeability in poly(ortho ester)s", *J.*

Membrance. Sci., vol. 65, pp. 269-275, 1992.
[http://dx.doi.org/10.1016/0376-7388(92)87028-V]

[11]   C.M. Hansen, and L. Just, "Water transport and condensation in fluoropolymer films", *Prog. Org. Coat.,* vol. 44, p. 259, 2002.
[http://dx.doi.org/10.1016/S0300-9440(02)00005-X]

[12]   J-S. Park, H. Chae, H.K. Chung, and S.I. Lee, "Thin film encapsulation for flexible AM-OLED: a review", *Semicond. Sci. Technol.,* vol. 26, p. 034001, 2011.
[http://dx.doi.org/10.1088/0268-1242/26/3/034001]

[13]   H. Lifka, H.A. van Esch, and J.J. Rosink, "Thin Film Encapsulation of OLED Displays with a NONON Stack", *SID 04 DIGEST,* pp. 1384-1387, 2004.

[14]   A.B. Chwang, M.A. Rothman, S.Y. Mao, R.H. Hewitt, M.S. Weaver, and J.A. Silvernail, "Thin film encapsulated flexible organic electroluminescent displays", *Appl. Phys. Lett.,* vol. 83, pp. 413-416, 2003.
[http://dx.doi.org/10.1063/1.1594284]

[15]   M.S. Weaver, L.A. Michalski, K. Rajan, M.A. Rothman, J.A. Silvernail, J.J. Brown, P.E. Burrows, G.L. Graff, M.E. Gross, P.M. Martin, M. Hall, E. Mast, C. Bonham, W. Bennett, and M. Zumhoff, "Organic light-emitting devices with extended operating lifetimes on plastic substrates", *Appl. Phys. Lett.,* vol. 81, pp. 2929-2932, 2002.
[http://dx.doi.org/10.1063/1.1514831]

[16]   C-S. Suen, and X. Chu, "Multilayer thin film barrier for protection of flex-electronics", *Solid State Technol,* vol. 51, pp. 36-39, 2008.

[17]   Y.C. Han, C. Jang, K.J. Kim, K.C. Choi, K. Jung, and B-S. Bae, "The encapsulation of an organic light-emitting diode using organic–inorganic hybrid materials and MgO", *Org. Electron.,* vol. 12, pp. 609-613, 2011.
[http://dx.doi.org/10.1016/j.orgel.2011.01.007]

[18]   K.H. Jung, J-Y. Bae, S.J. Park, S.H. Yoo, and B-S. Bae, "High performance organic-inorganic hybrid barrier coating for encapsulation of OLEDs", *J. Mater. Chem.,* vol. 21, pp. 1977-1983, 2011.
[http://dx.doi.org/10.1039/C0JM02008G]

[19]   J. Jin, J.J. Lee, B-S. Bae, S.J. Park, S. Yoo, and K. Jung, "Silica nanoparticle-embedded sol-gel organic/inorganic hybrid nanocomposite for transparent OLED encapsulation", *Org. Electron.,* vol. 13, pp. 53-57, 2012.
[http://dx.doi.org/10.1016/j.orgel.2011.09.008]

[20]   M. Leskela, and M. Ritala, "Atomic layer deposition (ALD): from precursors to thin film structures", *Thin Solid Films,* vol. 409, pp. 138-146, 2002.
[http://dx.doi.org/10.1016/S0040-6090(02)00117-7]

[21]   H. Kim, "Atomic layer deposition of metal and nitride thin films: Current research efforts and applications for semiconductor device processing", *J. Vac. Sci. Technol. B,* vol. 21, p. 2231, 2003.
[http://dx.doi.org/10.1116/1.1622676]

[22]   D.H. Levy, D. Freeman, S.F. Nelson, P.J. Cowdery-Corvan, and L.M. Irving, "Stable ZnO thin film transistors by fast open air atomic layer deposition", *Appl. Phys. Lett.,* vol. 92, p. 192101, 2008.

[http://dx.doi.org/10.1063/1.2924768]

[23]   S.J. Yun, Y-W. Ko, and J.W. Kim, "Passivation of organic light-emitting diodes with aluminum oxide thin films grown by plasma-enhanced atomic layer deposition", *Appl. Phys. Lett.,* vol. 85, pp. 4896-4898, 2004.
        [http://dx.doi.org/10.1063/1.1826238]

[24]   P.F. Carcia, R.S. McLean, M.H. Reilly, M.D. Groner, and S.M. George, "Ca test of Al_2O_3 gas diffusion barriers grown by atomic layer deposition on polymers", *Appl. Phys. Lett.,* vol. 89, p. 031915, 2006.
        [http://dx.doi.org/10.1063/1.2221912]

[25]   S-H. Park, J. Oh, C-S. Hwang, J-I. Lee, Y.S. Yang, and H.Y. Chu, "Ultrathin film encapsulation of an OLED by ALD", *Electrochem. Solid-State Lett.,* vol. 8, pp. H21-H23, 2005.
        [http://dx.doi.org/10.1149/1.1850396]

[26]   E. Langereis, M. Creatore, S.B. Heil, M.C. van de Sanden, and W.M. Kessels, "Plasma-assisted atomic layer deposition of Al_2O_3 moisture permeation barriers on polymers", *Appl. Phys. Lett.,* vol. 89, p. 081915, 2006.
        [http://dx.doi.org/10.1063/1.2338776]

[27]   W-S. Kim, M-G. Ko, T-S. Kim, S-K. Park, Y-K. Moon, S-H. Lee, J-G. Park, and J-W. Park, "Titanium dioxide thin films deposited by plasma enhanced atomic layer deposition for OLED passivation", *J. Nanosci. Nanotechnol.,* vol. 8, no. 9, pp. 4726-4729, 2008.
        [http://dx.doi.org/10.1166/jnn.2008.IC48] [PMID: 19049095]

[28]   A.A. Dameron, S.D. Davidson, B.B. Burton, P.F. Carcia, R.S. McLean, and S.M. George, "Gas diffusion barriers on polymers using multilayers fabricated by Al_2O_3 and rapid SiO_2 atomic layer deposition", *J. Phys. Chem. C,* vol. 112, pp. 4573-4580, 2008.
        [http://dx.doi.org/10.1021/jp076866+]

[29]   J. Meyer, D. Schneidenbach, T. Winkler, S. Hamwi, T. Weimann, P. Hinze, S. Ammermann, H-H. Johannes, T. Riedl, and W. Kowalsky, "Reliable thin film encapsulation for organic light emitting diodes grown by low-temperature atomic layer deposition", *Appl. Phys. Lett.,* vol. 94, p. 233305, 2009.
        [http://dx.doi.org/10.1063/1.3153123]

[30]   J. Meyer, P. Gorrn, F. Bertram, S. Hamwi, T. Winkler, H-H. Johannes, T. Weimann, P. Hinze, T. Riedl, and W. Kowalsky, "Al_2O_3/ZrO_2 nanolaminates as ultrahigh gas diffusion barriers—a strategy for reliable encapsulation of organic electronics", *Adv. Mater.,* vol. 21, pp. 1845-1849, 2009.
        [http://dx.doi.org/10.1002/adma.200803440]

CHAPTER 5

Thin Film Transistor Technology

Rongsheng Chen[*]

Department of Electronic and Computer Engineering, The Hong Kong University of Science and Technology, Hong Kong

Abstract: TFTs with large field-effect mobility, low threshold voltage, sharp sub-threshold swing, small leakage current, good uniformity, high stability and reliability are highly desired for high resolution active matrix OLED application. The three most popular technologies for the TFT backplane are a-Si TFT technology, LTPS technology and metal oxide TFT technology. There also exist some promising compounds TFTs, such as GaN, MoS_2 TFTs, and so on. The main properties of these technologies will be introduced and studied in this chapter.

Keywords: Amorphous silicon, Compound, Excimer laser crystallization, Gallium nitride, Metal induced crystallization, Metal oxide, Polycrystalline silicon, Solid phase crystallization, Thin film transistor.

INTRODUCTION

OLED displays can be classified as passive matrix OLED (PMOLED) and active matrix OLED (AMOLED) display.

PMOLED are particularly suitable for small area and low resolution display applications, such as mobile phones, MP3 players and automotive audio applications. The process for PMOLED is not complicated, which use separators or shadow mask [1]. For the driving scheme of PMOLED, an electric current is adjusted to pass through the selected pixels by scan line and simultaneously applying an electric current source to the corresponding data lines. There is no storage capacitor in the driving scheme of PMOLED, so the pixels are off most of the time during the driving. For the compensation, large current is needed to make OLED brighter. For example, if the display has 10 lines, one scan line that is on

---

[*] **Corresponding Author Rongsheng Chen:** Department of Electronic and Computer Engineering, The Hong Kong University of Science and Technology, Hong Kong; Tel.: +852-23588845; E-mail: rschen@connect.ust.hk

maybe 10 times brighter. So the efficiency and the lifetime are low for the PMOLED. PM is not a good driving method for high resolution OLED displays.

AM addressing is a better method to realize high resolution, high quality, and large size OLED display, as compared with PM addressing. The OLED works in the best power efficiency region and is driven by simple low voltage signals [2]. The AM TFT backplane acting as an array of switches controls the amplitude of current passing through each OLED pixel. Generally, this driving current is adjusted by at least two TFTs at each OLED pixel. One TFT is to turn on and turn off the charging of a storage capacitor C_S and the second TFT is to provide a constant current flow through the OLED pixel. The schematic of typical pixel circuit in AMOLED is shown in Fig. (**5.1**). This basic pixel circuit has two TFTs and one storage capacitor C_S. However, one TFT is required in AMLCD pixel. The switch TFT is acting as a switch while the driving TFT is providing the driving current required to turn on the OLED pixel. During the whole frame period, the driving TFT will still keep on providing current flow through the OLED pixel by the voltage held on the capacitor C_S. The main issue for this basic pixel driving circuit is its sensitivity to the non-uniformity of the TFT electrical performance over the whole backplane, which causes the non-uniformity of the luminance. In order to compensate the threshold voltage variation of TFTs, more complicated pixel circuits with additional TFTs have been designed and fabricated [3, 4].

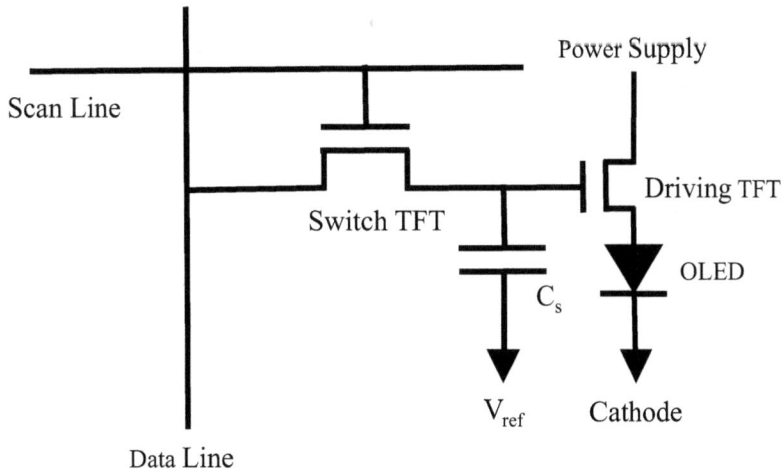

Fig. (5.1). Schematic diagram of the basic 2-TFTs pixel circuit in AMOLED.

For AMOLED, the peripheral circuit, such as control IC, scan driver, data driver, *etc.* are fully integrated and fabricated on the panel substrate and are called system-on-panel (SOP), as shown in Fig. (**5.2**). SOP is the flat panel display trend for high quality and high resolution display in the future. The cost of the display system is reduced dramatically by this technology because the peripheral ICs on silicon substrate can be eliminated. The connection of signal lines between active matrix panel and peripheral circuits become easy and more reliable [5]. The demands on TFT performance for SOP technology are much higher, as compared to that for application in AM driving pixel circuits.

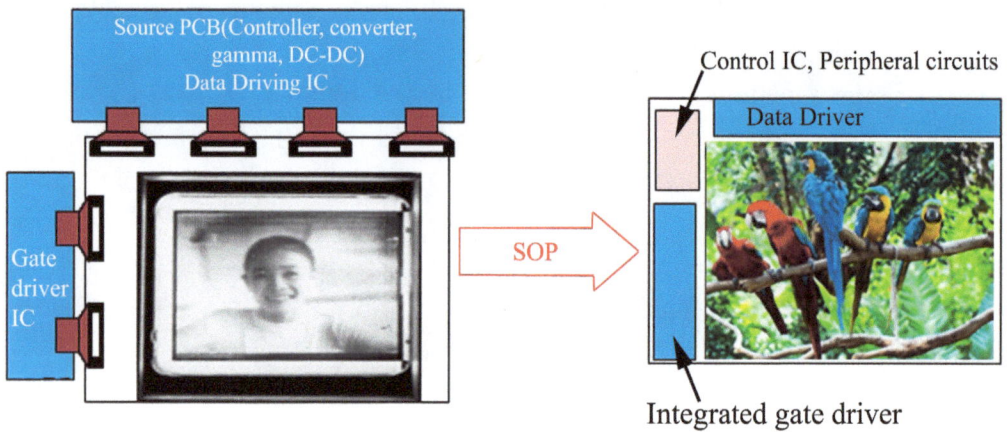

Fig. (5.2). The diagram of system-on-panel for AMOLED.

The TFTs on the glass for AM driving pixel circuits much show excellent electrical performance, such as high field-effect mobility (>20 cm²/Vs), small subthreshold swing (0.5 V/decade), low threshold voltage, high on/off current ratio, and low leakage current. And the uniformity of the TFTs performance in the whole panel is required to maintain the uniformity brightness of the AMOLED display. The conventional a-Si:H TFTs, which are widely used as switching TFT in AMLCD, have the advantage of good uniformity and low cost. However, their low field-effect mobility (<1 cm²/Vs) may be not sufficient to drive OLED pixels. The poly-Si TFTs are mainly used as switching and driving devices in the small size AMOLED display, due to their relative high mobility (> 50 cm²/Vs) and good electrical stability. However, due to the different grain size and grain boundaries in the poly-Si thin film, the poly-Si TFTs exhibit poor uniformity of the field-effect mobility and threshold voltage. Transparent amorphous metal oxide TFTs, such as a-IGZO TFTs have been used as driving devices in large-size AMOLED displays, because of their medium mobility and large area uniformity. However, the main issue for metal oxide TFTs is the poor electrical stability due to their sensitivity to oxygen and water vapor. Recently, several compound materials are investigated and studied for the active layer application for TFTs, such as nitride based compound and sulfide based compound. These compound TFTs can be potential candidates as driving devices in the next generation flat-panel displays, especially AMOLED display.

HISTORY OF THIN FILM TRANSISTORS

The history of thin-film transistors began with the research by P. K. Weimer at RCA laboratories in 1962 [6]. The TFT consisted of a semiconductor layer of cadmium sulfide (CdS) and a gate insulator silicon monoxide in a staggered structure, as shown in Fig. (**5.3**). A transconductance of 25000 $\mu A/V$ with a gate capacitance value of about 50 pF and oscillation frequency about 20 MHz were obtained. Later, Cadmium selenide (CdSe) was used as active layer to produce TFTs with similar results by F. V. Shallcross [7]. Two years later, P. K. Weimer developed p-type TFTs using tellurium as the active channel layer [8].

The research activity of TFT technology become to decline at the end of the 1960s because of the emergence of the insulated-gate field-effect transistor (IGFET) in 1962 [9], with the advantage of more dense integrated circuit chips and low cost. The great opportunity for TFTs technology research appeared again when people realized that low cost is inseparable from small chip size for crystalline silicon and large arrays of low cost electronics are required for some application, such as AMLCD. AMLCD with CdSe TFTs as active matrix had been demonstrated by T. P. Brody in 1973 [10, 11]. The poor DC stability of CdSe TFTs was a main issue preventing application in large size display and commercialization [12].

(a)

(b)

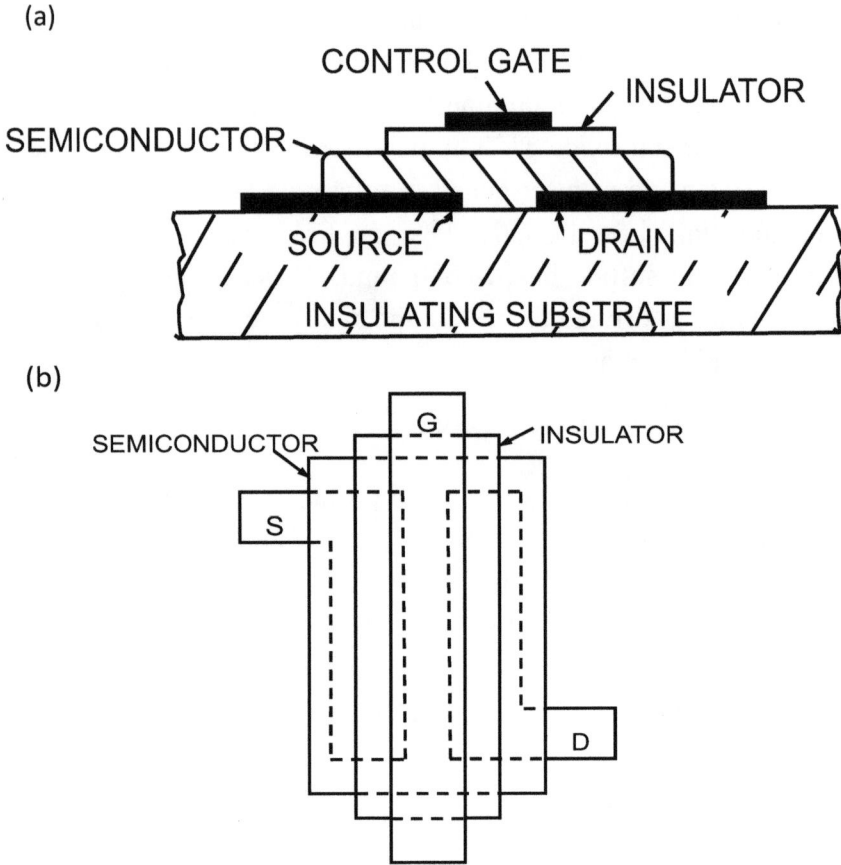

Fig. (5.3). (**a**) The cross sectional schematic and (**b**) top view of the CdS TFT [6].

HYDROGENATED AMORPHOUS SILICON TFT TECHNOLOGY

The technology of a-Si:H TFT played an important role in the AMLCD development. AMLCD, which replaced the CRT display in the market, dominate flat panel display market in these ten years, such as LCD-TV, LCD computer monitor, *etc*. The breakthrough of a-Si:H TFT technology is the discovery that the hydrogenated amorphous silicon could exhibit n-type or p-type conductivity depending on the donors or acceptors dopant [13]. The mix gases of SiH_4+PH_3 and $SiH_4+B_2H_6$ were used to deposit n-type and p-type material in plasma enhanced chemical vapor deposition (PECVD) process, respectively. The TFTs using a-Si as the active channel layer was developed and fabricated in 1979 [14].

The schematic diagram is shown in Fig. (**5.4**). A-Si:H TFT became attractive research topic in the 1980s with the application in active matrix of LCD panel. Lots of research had been performed to achieve a good performance a-Si:H TFT, including low off-current (<1 pA), high on/off current ratio (>10^7), reasonable mobility (0.5-1.0 cm^2/Vs) and good stability.

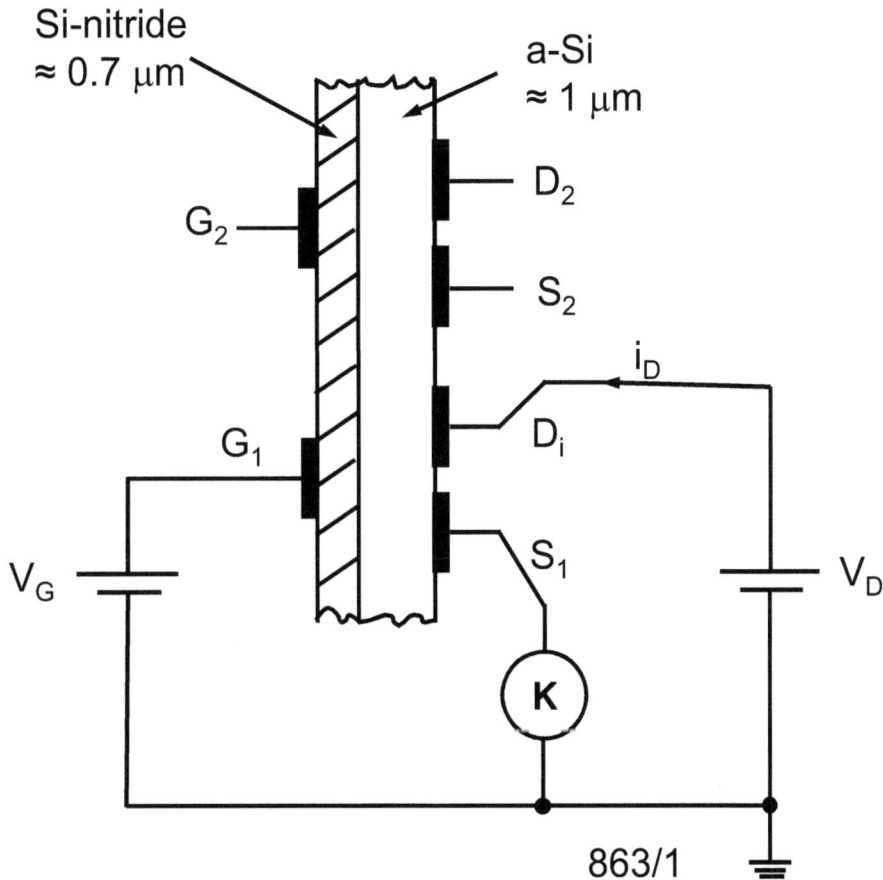

Fig. (5.4). Schematic diagram illustrating the design of the a-Si TFT [14].

LOW TEMPERATURE POLYCRYSTALLLINE SILICON TFT TECHNOLOGY

Low temperature polycrystalline silicon (LTPS) TFTs exhibits much larger mobility than a-Si TFTs. The poly-Si film can be directly deposited in the polycrystalline phase, or recrystallized from the amorphous phase through different annealing schemes. The mobility of LTPS TFTs varies from several tens to several hundred cm^2/Vs, which fulfills the requirement of high resolution AMLCD and AMOLED displays with high pixel density. Multiplexers for data drivers and shift registers for gate drivers can also be built using high performance LTPS TFTs. Moreover, LTPS TFTs demonstrate superior stability.

Advanced laser annealing techniques produces poly-Si with high crystallinity [15]. Other crystallization technologies, such as metal induced crystallization (MIC) and solid phase crystallization (SPC), have also been extensively studied. Laser annealing equipment is expensive, and the size of the substrate is limited. Moreover, laser annealed poly-Si TFTs shown large variations in TFT performance because of the relatively large grain size and randomly distributed grain boundaries. In order to improve the light intensity uniformity, complicated compensation pixel circuits should be used. The compensation circuit usually consists of 4-6 TFTs, which will decrease the aperture ratio of the pixel. Using small grain LTPS material helps improve the uniformity, but the TFT performance will be sacrificed to some degree.

SPC Technology

SPC refers to the conversion of a-Si to the polycrystalline phase by thermal annealing. In the SPC process, the phase transition is conducted directly from one solid phase (amorphous) to another solid phase (polycrystalline phase). No melting of a-Si happens during the SPC process.

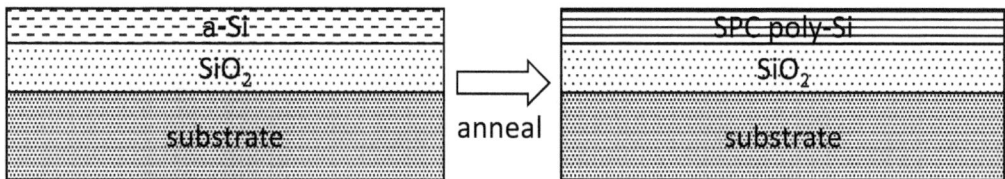

Fig. (5.5). Schematic of SPC process.

The conventional SPC process is simple and easy to obtain poly-Si. No mask or catalyst is involved in this process. After a-Si deposition, the substrate is directly sent into a furnace to anneal at a specific temperature for a certain time. Thermal energy is the only driving force to convert a-Si into poly-Si. The schematics of this SPC process are shown in Fig. (**5.5**). In an early publication of an SPC TFT [16], 550 °C annealing for 72 hours allowed full crystallization of 50 nm-thick LPCVD a-Si film. The TFT based on this SPC film showed field effect mobility of 19.2 cm^2/Vs and threshold voltage of 1V.

In the SPC process, nuclei density is high at or near the interface between a-Si and the substrate [17]. As a result, crystallization starts here. Microtwins and stacking faults defects usually exist because of a large magnitude of tensile stress at or near the interface between a-Si and SiO_2 substrate. The grains prefer to grow to an elliptical shape. Formation of grain boundaries is due to an orientation mismatch between the nuclei. A high density of orientation mismatched nuclei result in small grains and a high grain boundary density. The key factor of the SPC process is the nuclei density, which is affected by the a-Si deposition method and condition. Nuclei are necessary to induce crystallization. A higher density of nuclei results in a fast crystallization rate but small grain size. Interface engineering is necessary to control the nuclei density to produce the desired crystallization rate and quality of grain. A well-known method is silicon self-implantation [18, 19]. Silicon film is firstly deposited in the poly-Si or a-Si phase by CVD to offer a sufficient amount of pre-existing sub-grain. The film is then amorphized by silicon implantation. As a result, seeds with <110> orientation survive due to the channeling effect. Other seeds are sufficiently amorphized, resulting in reduced density of nuclei with optimized orientation. Larger grain size can be obtained through this method. Field effect mobility of the SPC TFT is about 37-60 cm^2/Vs at optimized experiment conditions. A CMOS ring oscillator has also been demonstrated using this type of TFT [19]. The drawback of this method is that the crystallization time is very long, which is a direct result of reduced nuclei density. Other interface-nucleation suppression methods include argon implantation [20], oxygen blowing at the initial a-Si deposition period [21] and so on. However, large grain size and fast crystallization cannot be realized at the same time, which is due to their opposite requirements for nuclei density.

Rapid thermal annealing (RTA) is capable of heating up and cooling down the substrate in a very short period of time. Glass substrates can be exposed to temperatures above their strain temperature for a very short period of time [22].

Crystallization time of 55nm thick a-Si is as short as 40s at 750 °C, as reported by Bonnel [23] in the early 1990s. TFTs fabricated through RTA exhibit a field effect mobility of about 20 cm^2/Vs. RTA and the furnace SPC process can be applied in combination. A high-temperature RTA nucleation step was proposed to grow large size grain poly-Si thin film by Subramanian [24]. Quenching from RTA to low temperature furnace annealing retards the nucleation and enhances the grain size.

A magnetic field can be externally applied to accelerate the SPC process [25]. Alternating magnetic fields to assist crystallization which has been named advanced solid phase crystallization (a-SPC) was proposed by LG Display [26]. An alternating magnetic field of 500 G, 230 KHz is applied in a perpendicular direction to the a-Si film. Crystallization can be achieved at 430 °C annealing for 8 minutes. It is believed that the thermally generated carriers can generate thermal energy when moving around the magnetic field; this Joule heating would accelerate the crystallization rate of the a-Si. Grain size is estimated to be around 250 nm. With the channel lightly doped with boron before the annealing and the H_2 plasma passivation, TFTs using this field aided crystallized film as an active layer with 100 nm SiO_2 as gate insulator exhibit a mobility of 20 cm^2/Vs, V_{th} of -3.5V and an SS value of 0.7 V/decade [27] . In terms of uniformity, the output current uniformity is ±7%, which is much better than that of the laser crystallized ones (uniformity is ±20% as they reported). Based on the magnetic field assisted crystallization technology, a 15-inch bottom emitting XGA AMOLED display panel which shows good uniformity (9.5%) in terms of luminance is demonstrated using p-type TFTs with 4T1C pixel and symmetric driving method without V_{th} compensation [28].

MIC Technology

Metal catalysts can be introduced into a-Si to lower the activation energy of crystal-lization, resulting in a shorter process time or lower reaction temperature. Crystallization with a metal catalyst is categorized as a MIC process. The metal atoms either form eutectics/alloy with silicon or form metal silicide. Au, Al, Sb and In tend to form eutectics/alloys with silicon [29]. The nucleation of a-Si and growth of crystalline silicon may be enhanced by the involvement of metal atoms

[30]. Pb, Ti and Ni form silicide with silicon [29]. Nickel is most widely applied in the MIC process. $NiSi_2$ precipitates can be formed within 15 min at 400 °C [31]. $NiSi_2$ is of cubic crystal structure with the lattice constant equal to 5.406Å .This value is close to that of c-Si, which is equal to 5.430 Å. $NiSi_2$ occurs irregularly within the a-Si thin film as nuclei, the orientation of which determines the orientations of the crystalline structure. The earliest stage of the a-Si/c-Si transformation appears to involve the formation of a few layers of crystalline silicon . The Ni atoms would diffuse to the growing interface from the rear of $NiSi_2$ precipitate. Different variations of the MIC process have been proposed based on controlling the Ni diffusion and grain growth. Some of the well-known processes are reviewed below.

The simplest method of MIC is to deposit nickel uniformly on the a-Si layer, as shown in Fig. (**5.6**). Ni can be sputtered or evaporated onto the film. Immersion of the substrate into a Ni containing solution [32] has also been reported for MIC. The substrate with Ni is then annealed in a furnace or RTA equipment [33]. $NiSi_2$ is first formed as the nuclei. The main direction of Ni diffusion and grain growth is along the depth of the a-Si film. After annealing, the unreacted Ni is removed. The MIC poly-Si thin film is then patterned into active islands for the subsequent TFT fabrication processes. The crystallization rate of MIC is much faster than for SPC. The grain growth rate and size of MIC poly-Si is dependent on the amount of Ni and the annealing temperature. Generally speaking, the grain size of direct MIC is the smallest among all MIC methods, but still larger than those using the conventional SPC process.

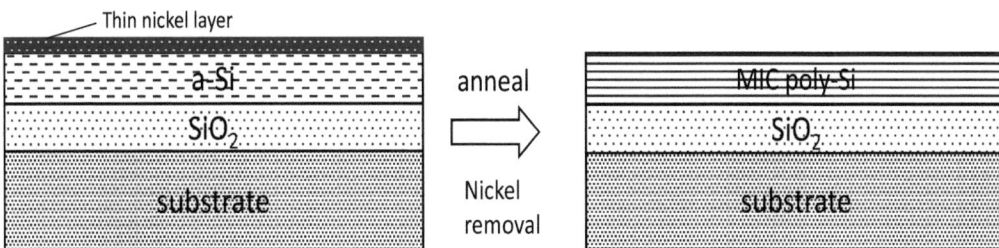

Fig. (5.6). Schematic direct MIC process.

Electric field aided MIC, where the MIC process is conducted under an electric field, has been proposed by Jin Jang's group [34 - 37]. The rate of MIC could be enhanced by an electric field. 400 nm thick a-Si could be crystallized at 500 °C for 600 seconds in the presence of an 80 V/cm electric field. The Ni atoms in the $NiSi_2$ are then negatively charged and the diffusion coefficient of Ni is enhanced in the presence of the electric field, resulting in a much faster crystallization rate. MIC TFTs fabricated based on field aided MIC have exhibited a field effect mobility of 58 cm^2/Vs and a threshold voltage of -1.8V.

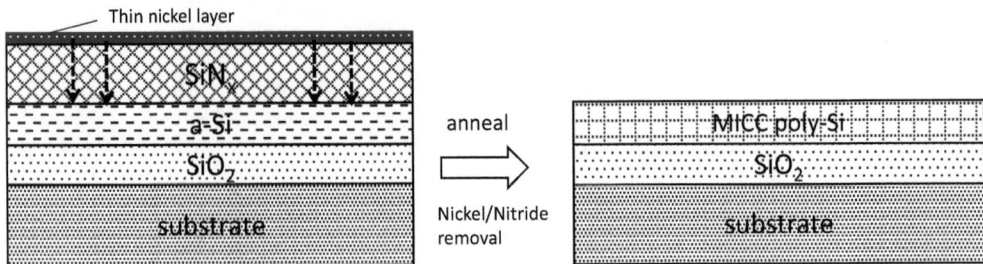

Fig. (5.7). Schematic MICC process.

Metal Induced Crystallization through a Cap Layer (MICC) has been proposed to enlarge the grain size of poly-Si thin film [38, 39]. As shown in Fig. (**5.7**), a layer of SiN_x is deposited on the a-Si layer. The SiN_x layer acts as a filter of Ni. Annealing can be done through furnace annealing or RTA. Ni diffuses slowly through the cap layer and forms $NiSi_2$ with a-Si.

Due to reduced Ni density, the grain size is greatly enlarged. The grain size of the poly-Si obtained by using a 0.5 nm Ni layer is 20-90 µm and its surface roughness is better than by using direct MIC methods. Field aided MICC has also been reported [40]. The p-type MICC poly-Si TFT [41] showed an extracted field-effect mobility of 101 cm^2/Vs and a leakage current of 10^{-12} A/µm at $V_{ds} = -10$ V with good stability, which is due to the smooth surface of the active MICC poly-Si layer.

The MIC process can also be done with patterned induced windows. Ni is directly in contact with the a-Si only inside the induced windows. MIC with induced windows can be further divided into metal induced lateral crystallization (MILC), MIUC and so on. MILC and MIUC require certain masks to selectively expose the a-Si region to the nickel.

The MILC process was first proposed by Lee and Joo [42, 43] in 1996. As shown in Fig. (**5.8**), the self-aligned gate electrodes are used as the mask. Nickel is sputtered after the gate electrode formation and contacted directly with the a-Si in the source and drain region. The a-Si in the source and drain region react with the nickel and crystallize through the same mechanism as the direct MIC process. Crystallization then laterally extends into the channel region until the two fronts merge. A major grain boundary is created at the center of the channel. 500°C annealing in the mixture of N_2 and H_2 allows a lateral crystallization rate of 1.6 µm/hour. The metal concentration in the MILC region is lower compared with the MIC region. The Ni-MILC process can produce large crystalline grains because of the preferred <110> oriented growth direction of $NiSi_2$.

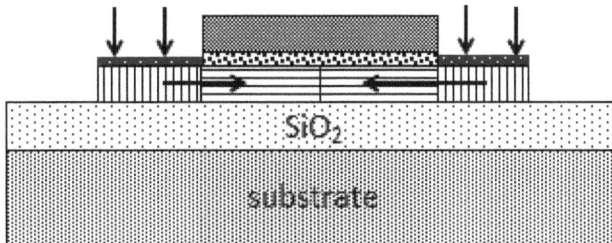

Fig. (5.8). Schematic MILC process.

MILC TFTs show high mobility and lower V_{th}, due to their larger grain size and fewer intra grain micro defects. N-channel poly-Si TFTs fabricated at 500 °C by MILC exhibit a field-effect mobility of 121 cm^2/Vs, a V_{th} of 1.2 V, and an on-off current ratio of higher than 10^6. Compared with direct MIC methods, MILC technology gives larger grain size and lower density in grain boundaries and defects in the lateral growth region, without introducing any additional mask or photolithography process.

The MILC rate can be enhanced under an electric field [44]. The electrons move freely to the a-Si/silicide interface and pile up. The nickel diffusion is accelerated because of the local electric field. In addition to Ni, Pd is also found to be able to induce lateral crystallization. But Pd MILC is not suitable for TFTs due to the large micro-twin defect density in poly-Si thin film.

However, researchers have determined that the interface between MIC and MILC has many defects in the grain boundary [45]. These defects negatively affect the V_{th} and SS of the MILC TFT. Moreover, with the charged defect states, the electric field and impact ionization rate increase at the drain junction, resulting in the kink effect for MILC TFTs. An offset structure which is known as MIUC has also been proposed to remove the MIC/MILC interface from source/drain region [46 - 48].

The schematic diagram of the MIUC process is shown in Fig. (**5.9**). An additional mask is applied to pattern some windows on the LTO layer above the a-Si layer. Nickel is deposited onto the substrate and comes into contact with the a-Si only inside the windows. During annealing, MIC starts from the a-Si inside the windows and laterally extends until it meets another crystallization front. After crystallization, nickel residue and the LTO layer are removed. The crystallized poly-Si layer is then ready for active layer patterning. The active island is placed where there is no major grain boundary or MIC/MILC interface inside the channel. Compared with MILC devices, MIUC TFTs have shown superior results in terms of almost all parameters, such as μ_{FE}, V_{th}, SS, leakage current, on-off ratio and uniformity.

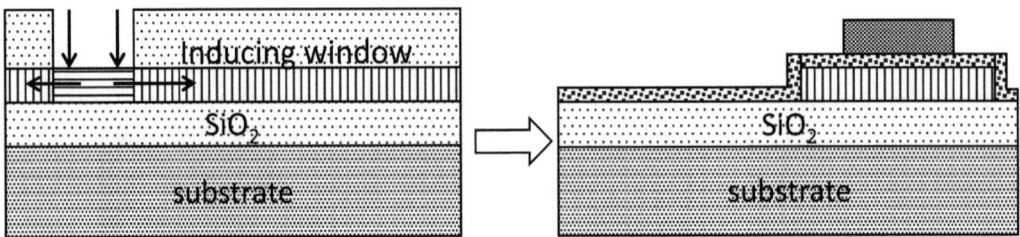

Fig. (5.9). Schematic MIUC process.

ELA Technology

Laser crystallization gives the poly-Si film with best quality. Excimer laser annealing (ELA) crystallization is one of the most applied and studied laser crystallization techniques. The ultraviolet output of excimer lasers has wavelengths of 308 nm, 248 nm, and 193 nm for XeCl, KrF, and ArF gas mixtures, respectively. The pulse has a very short duration time between 10-30 ns. The a-Si layer absorbs the UV light and has high temperature at the surface because of the strong absorption of silicon [49]. As a result, local melting of a-Si happens, while the substrate has no significant heating. Therefore, the ELA process is compatible with glass substrates, which is suitable for large size flat panel display. The poly-Si grain is crystallized from the melting phase with high crystallinity [50]. The crystallization process is sensitive to experiment conditions such as the laser intensity, irradiation dose, spot size and so on [51]. The grain size can be enhanced by doing substrate heating and laser irradiation simultaneously [52]. The drawback of the laser annealing process is the high cost of the equipment. Moreover, the scanning process is slow when the spot size is small with large size substrates.

Both poly-Si and a-Si:H TFTs with bottom gate structure have been fabricated on the same glass substrate [53]. The laser is scanned selectively in the area prepared for the driving circuits to save the process time. This method is suitable for LCDs with driving circuits integrated on panel. Combinations of different crystallization methods have also been proposed by different groups. ELA of SPC films [54], ELA of MILC films [15] and so on have been reported to be able to further improve the quality of SPC/MILC films and the resulting TFTs performed better than those based on ELA of a-Si.

Sequential lateral solidification (SLS) is a laser crystallization technology that has advantage of large grain size and controllable of grain boundary location in the poly-Si film [55 - 57]. Phase modulated ELA has also been proposed to produce position-controlled large-grains. The phase of excimer-laser light can be modulated with a phase-shift mask between the substrate and the light source [58]. New laser sources, such as a diode pumped solid state continuous wave laser [59, 60], have also been proposed as alterative laser crystallization methods.

However, there is always a compromise between the electrical performance and the uniformity of device characteristics [61]. Moreover, the high cost and low throughput limits the application of laser crystallized poly-Si in larger area display panels with low cost.

Bridge Grain Technology

Improvement in the poly-Si thin film quality through an enlargement of the average grain size or a decrease of the intra-grain defect will help reduce the leakage current at low electric fields because of the decrease of trap states. Meanwhile, with high electric fields at the drain region, the leakage current is mainly due to the band-to-band tunneling. An electric field relief structure is necessary for either large or small grain TFTs.

The characteristics of poly-Si TFTs depend not only on the grain size but also the location of grain-boundaries. To improve the device characteristics uniformity, grain boundary location control is of great importance.

Enlarging the grain size or scaling down the TFT dimensions will help improve the characteristics of TFTs, but will also degrade the device uniformity. Scaling down the devices to submicrometer scale will also result in a seriously degraded yield. A small grain size results in uniform performance, but the performance is uniformly inferior.

There have been many attempts to control the grain size and grain boundary locations [62]; however, the nucleation of the crystallization process is always with some randomness. Moreover, intra-grain defects also affect the performance of TFTs.

Bridged-grain (BG) technology is developed and applied in small grain polycrystalline silicon TFTs, such as SPC and MIC TFTs. Application of BG technology greatly enhances the performance and the uniformity of these small grain TFTs.

The schematic diagram of top-view and cross-sectional view of a BG TFT structure are shown in Fig. (**5.10**) [63]. There have several BG regions (BG lines), which is heavily doped along the channel. The BG lines are perpendicular to the direction of the source/drain current. The dopant of BG lines is the same as the source and drain regions. The width of BG lines is usually from 0.5 μm to 2 μm depending on the process and design requirement. Short channel effects (SCEs) occur in the BG TFTs depending on the width and spacing of the BG lines.

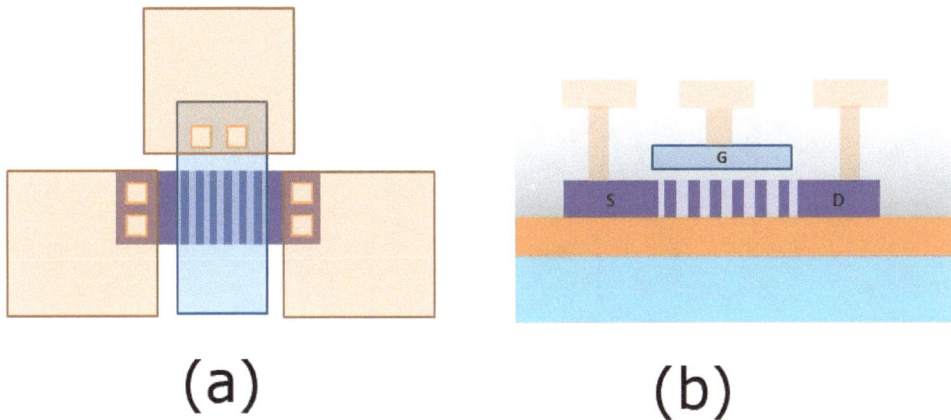

Fig. (5.10). (a) Top- and (b) Cross-sectional schematic views of a BG TFT [63].

The benefits of employing a BG structure originate from several aspects: (1)When the device is on, the V_{th} will reduce and the carrier mobility will increase due to the SCEs effects. (2) When the device is off, the leakage current can be dramatically mitigated by the multi-junction effects [64] that exist in the BG structure. (3) When the grain size is made comparable to the channel length, the defect density in the channel can be reduced and the mobility and SS can be improved.

The path of the current flow in BG SPC poly-Si channel is shown in Fig. (**5.11**). The grains with different shape and size in the poly-Si are labeled with different color in the diagram. The rectangular in blue regions are the BG lines, which are heavily doped. It is conductive in the BG line. So the channel in TFT consists of several intrinsic regions and heavily doped regions. The current in the channel will flow with fewer grain boundaries and lower barrier potential in the intrinsic region and then flow into next BG region, as shown in Fig. (**5.11**). The total grain boundaries and the barrier potential for the current path in the channel will be reduced, which will lead to the higher mobility, lower threshold voltage, larger on current and steeper subthreshold behavior for the BG TFT.

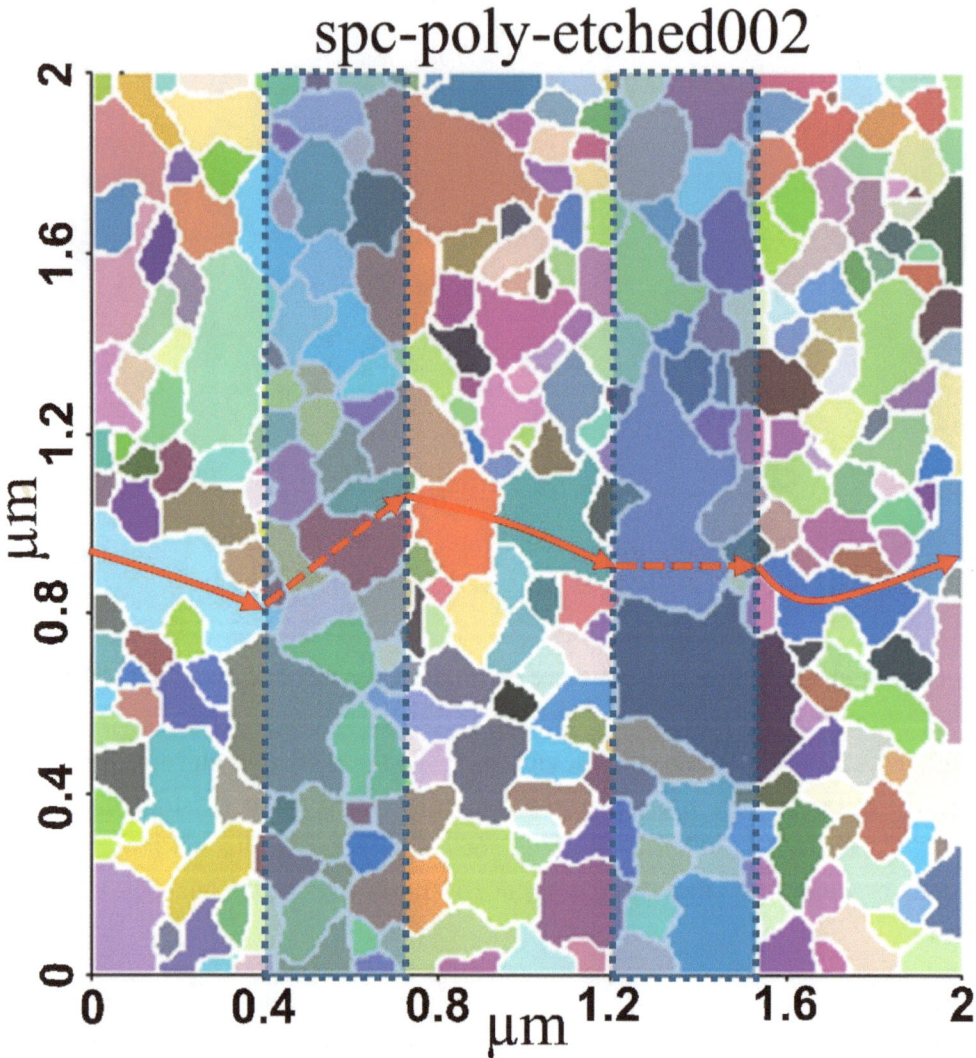

Fig. (5.11). The current flow diagram in BG poly-Si thin film [63].

Furthermore, there exists multi-junction effect in the BG structure. There are several floating conductive regions in the channel of a BG TFT, which is similar to the split-gate TFTs [64]. The effective V_{ds} of the subchannel is reduced greatly and the hot carrier effects at the drain region can also be mitigated due to the lower electric field. In terms of the leakage current, the multi-junction structure effectively reduces the electric field near the drain region. This effect is verified by the simulation results of electric field distribution along the channel under the off-state for regular single gate TFTs (L=6 μm) and BG TFTs (L=12 μm), as shown in Fig. (**5.12**).

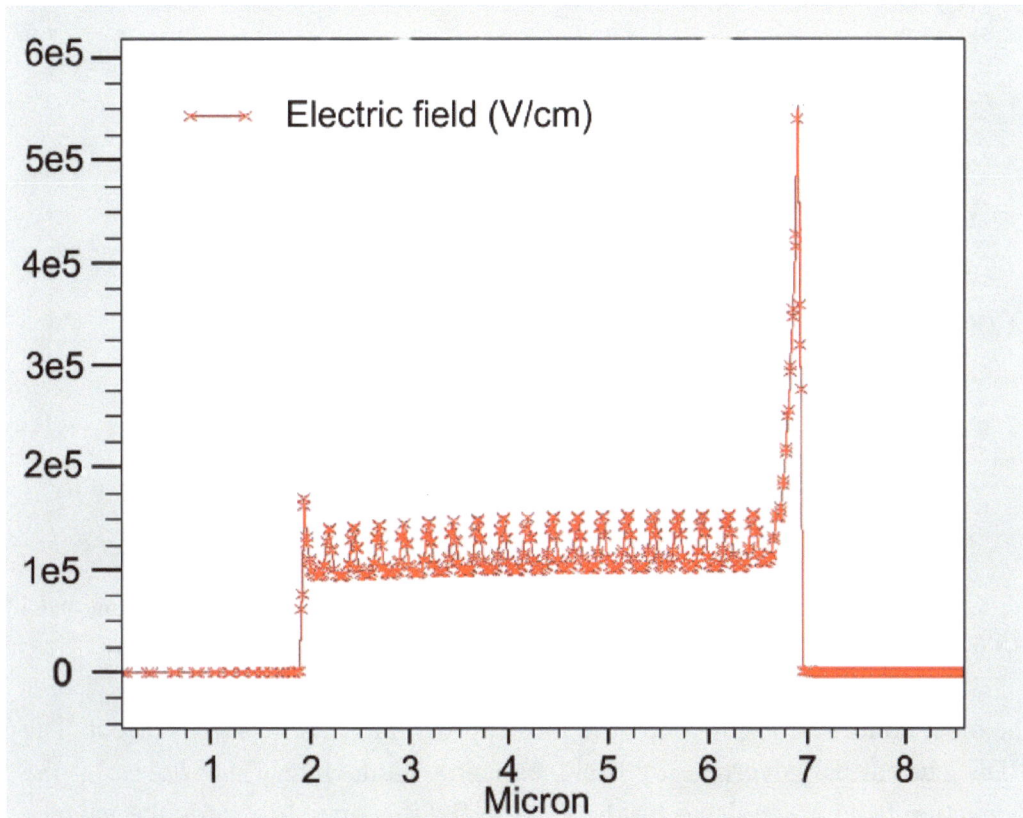

Fig. (5.12). Simulation results of electric field distribution along the channel under the off-state of (**a**) single gate TFT and (**b**) BG TFT.

The transfer characteristics of the conventional typical SPC TFTs and the BG-SPC TFTs with V_{ds}= (0.1 V and (5.1 V are shown in Fig. (**5.13**). The inset of Fig. (**5.13**) shows the G_m characteristic with V_{ds}= (0.1 V. As compared to the conventional typical SPC TFTs, the BG SPC TFTs showed much better electrical performance, such as lower threshold voltage, steeper subthreshold slope, higher on-off current ratio, and higher field-effect mobility.

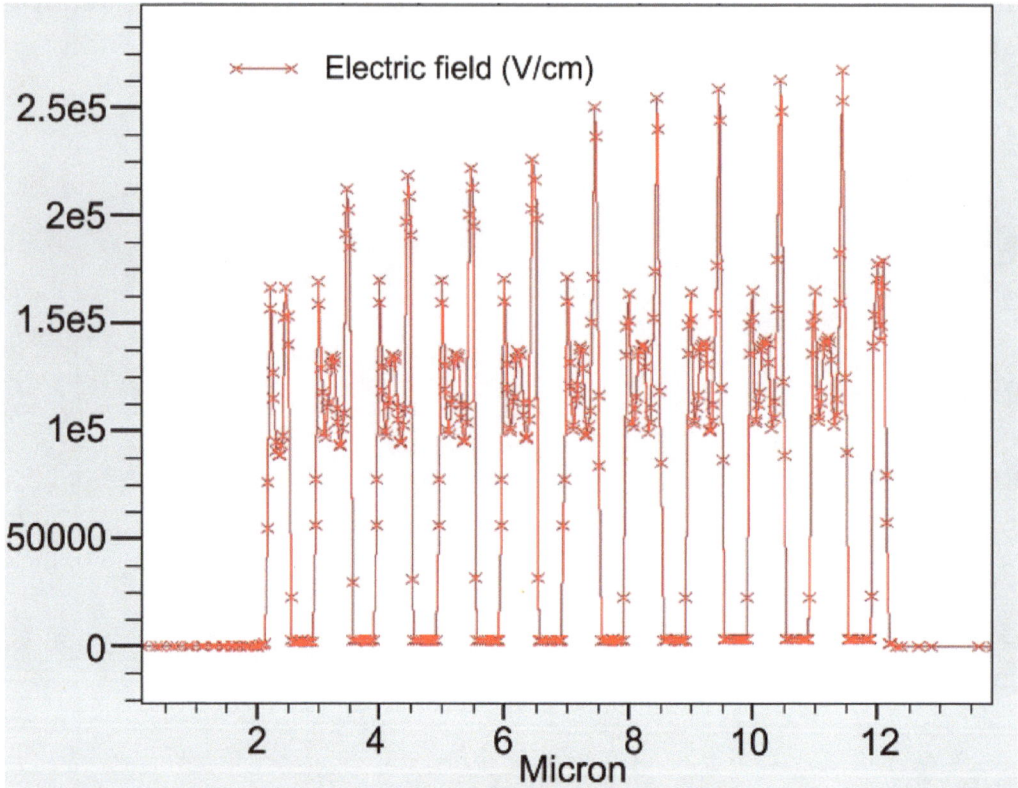

Fig. (5.13). Comparison of the transfer characteristics of the SPC and BG-SPC TFTs (W/L=24μm/10μm) [63].

The most noticeable improvement is the suppression of the leakage current. The GIDL current is governed by field enhanced tunneling. The holes in the conduction band are assisted by the electric field to tunnel towards the valence band *via* the grain boundaries traps, whereas the leakage current I_{min} is mainly induced by the thermally generated carriers *via* the trap states [65].

Compared with the SPC devices, the GIDL of the BG-SPC devices is suppressed by more than 145 times. I_{min} is decreased by 8.6 times. With an increasing number of BG lines inside the channel, the electric field, and therefore GIDL, is further reduced, as shown in Fig. (**5.14**).

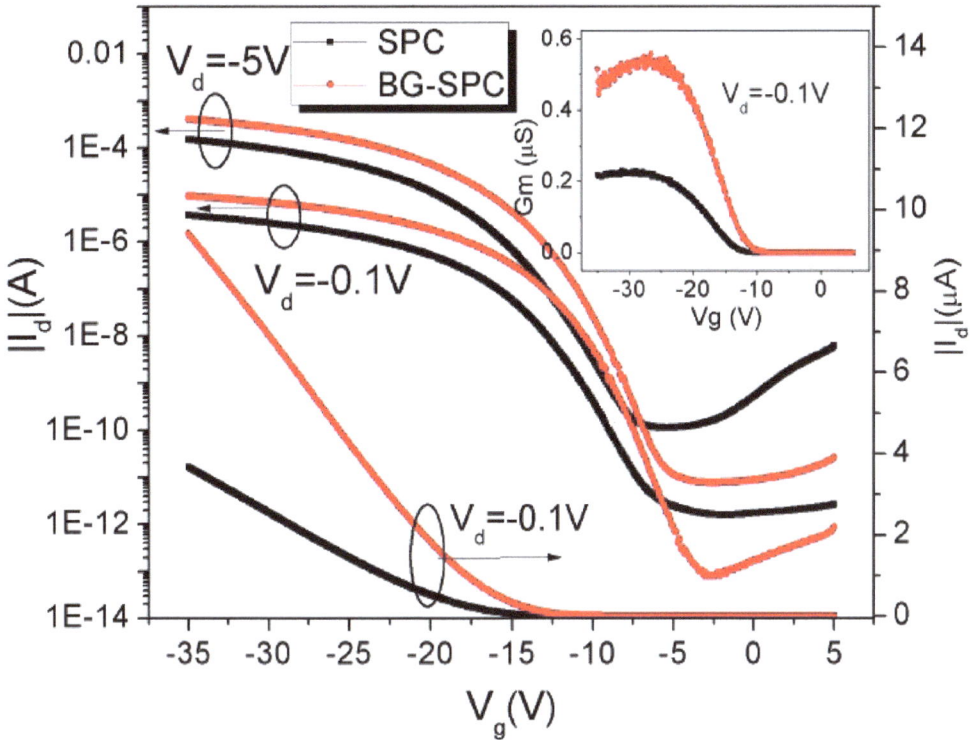

Fig. (5.14). GIDL of SPC/BG-SPC TFTs with different L [63].

METAL OXIDE SEMICONDUCTOR TFTS

Zinc Oxide TFTs

Recently, oxide semiconductor is a promising channel material in TFT for flat panel display applications, especially in AMOLED display [66], challenging conventional silicon not only in conventional applications but also exploring novel areas, such as paper electronics [67].

Among various oxide semiconductors, ZnO is one of the most widely studied. It is not only used as a transparent conducting oxide (TCO) material, but also in health care application due to its absorption of the UV light. The wide band gap for ZnO is from 3.1 to 3.37 eV. It is direct band gap material. The electron and hole mobility are reported to be about 200 cm^2/Vs and 5-50 cm^2/Vs, respectively [68]. ZnO have a wurtzite structure with a preferred c-axis orientation in its polycrystalline structure. ZnO thin films can be deposited by many techniques, such as reactive sputtering (DC [69], RF [70, 71], ion beam [72]), activated reactive evaporation (ARE) [73], pulse laser deposition (PLD), metal-organic chemical vapor deposition (MOCVD), and electrochemical reaction.

In 2003, Hoffman *et al.* developed and fabricated ZnO TFTs using ion beam sputtering technique on glass. A 200 nm thick ITO layer was used as gate electrode deposited by sputtering technique and a 220 nm thick aluminum-titanium oxide was used as gate dielectric deposited by atomic layer deposition [74]. The typical bottom-gate structure is shown in Fig. (**5.15**). The channel layer ZnO thin film was deposited by sputtering technique at room temperature. After that, sample was transferred to rapid thermal anneal process at 700 °C in oxygen ambient to decrease the oxygen vacancy and improve the ZnO thin film quality. The extracted field effect channel mobility of the device was 2.5 cm^2/Vs. The threshold voltage of the TFT devices ranges from 10 to 20 V, and an on/off current ratio is about 10^7. The optical transmission of the proposed ZnO TFTs with the substrate is about 75% in the visible range, as shown in Fig. (**5.16**).

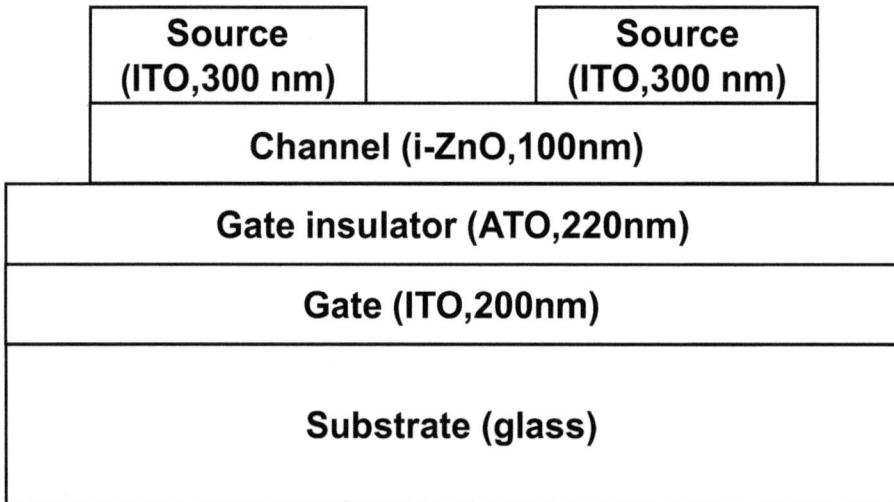

Source (ITO,300 nm)		Source (ITO,300 nm)
Channel (i-ZnO,100nm)		
Gate insulator (ATO,220nm)		
Gate (ITO,200nm)		
Substrate (glass)		

Fig. (5.15). The bottom-gate structure ZnO TFTs proposed by Hoffman *et al.* [74].

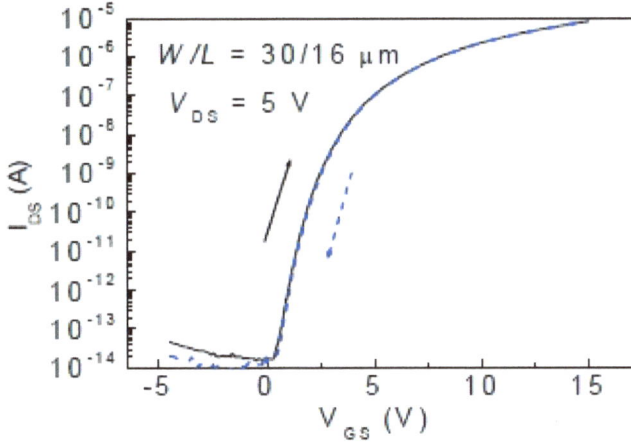

Fig. (5.16). Optical transmittance spectrum for the ZnO TFTs including substrate [74].

In the same year, Bottom-gate type ZnO TFTs with the active channel layer ZnO deposited by PLD at 450 °C were demonstrated by Masuda *et al.* [75]. The gate leakage current was reduced utilizing SiO_2 and Si_3N_4 double layer gate dielectric. Then, ZnO thin film was directly deposited on top of the silicon nitride. The Si_3N_4 could prevent the Zn diffusion from the active channel into the SiO_2. But the extracted field effect mobility obtained was quite low, about 0.97 cm²/*Vs*. The optical transmittance of ZnO TFTs with glass substrate was more than 80%, which can be applied to transparent electronics.

In 2003, Carcia *et al.* fabricated bottom-gate type ZnO TFTs with heavily doped n-type silicon substrates as gate electrode. A 100 nm thick thermal oxide layer as gate dielectric [76]. The active layer ZnO thin film was deposited by rf magnetron sputtering technique at near room temperature. The ZnO TFTs exhibit field-effect mobility of 2 cm²/*Vs* and an on/off current ratio of larger than 10^6.

Additionally in 2003, Nishi *et al.* developed bottom-gate type ZnO TFTs with improved interface between ZnO channel layer and Si_3N_4 gate insulator. $CaHfO_x$ buffer layer was introduced at the interface [77]. The channel layer ZnO is deposited by PLD at 300 °C with aluminum for the source and drain contacts. The field-effect mobility can reach about 7 cm²/*Vs*, due to the introduction of this buffer layer. The ZnO TFTs also have high on/off current ratio, about 10^6. Because of the relative good electrical performance, the proposed ZnO TFTs are competitive with those commercial a-Si TFTs used in active-matrix displays. The process temperature for ZnO thin film deposition could be as low as 150 °C with no obvious degradation for device electrical performance.

In the same year, spin-coating method was applied for the deposition of ZnO thin film,which was used as active channel layer to produce ZnO TFTs by Norris *et al.* [78]. The devices have a staggered bottom-gate structure with superlattice of Al_2O_3 and TiO_2 (ATO) as the gate insulator. Annealing process at 600 °C for 10 min was performed to increase the channel resistance, and then rapid thermal annealing at 700 °C was also carried out to improve the crystallinity of the ZnO thin film. For the purpose of transparency, ITO was deposited by ion-beam sputtering to act as source/drain electrode. The devices showed field-effect mobility of 0.2 cm^2Vs. But the device exhibited high sensitivity to light due to the poor surface roughness.

In 2004, Fortunato *et al.* demonstrated relative high performance bottom-gate configuration ZnO TFTs. The channel layer ZnO was deposited by rf magnetron sputtering processed at room temperature [79]. ATO is utilized as the gate dielectric and gallium-doped zinc oxide is deposited as the source and drain electrodes. The devices showed a saturation mobility of 27 cm^2/Vs, an on/off current ratio of 3×10^5, and a threshold voltage of 19 V, which is quite large.

Amorphous Oxide Semiconductors and TFTs

Mobility is the most crucial parameter for the TFT device characteristics in the flat panel display application. The ZnO TFTs were well studied with the field effect mobility ranging from 1 to 50 cm^2/Vs depending on the microstructure of the ZnO, thermal annealing process, gate dielectric material, and so on [74,75,79]. However, the main issue preventing from commercialization for ZnO TFTs is the non-uniformity of the electrical performance due to the grain size and grain boundaries defects in the ZnO thin film. .

So amorphous state of oxide semiconductor with low process temperature, high electron mobility, and smooth surface is necessary. The uniformity is improved greatly due to the fact that there is no grain and grain boundary in the thin film. The main issues are the ability to deposit the amorphous semiconductors at low process temperatures, which is suitable for application of flexible flat panel display. This novel type semiconductor material, amorphous oxides semiconductors (AOS) represents a novel idea and has advantages of high transparency, low process temperature, high electron mobility, high uniformity and easy fabrication. AOS do not have grain boundaries inside the film, which has high uniformity and is a great advantage for large area application.

Amorphous oxides have heavy post transition metal cations with $(n-1)d^{10}ns^{0}$ ($n \geq 4$) electronic configurations reserving relatively high electron mobility despite their amorphous state [80, 81]. The comparison diagram in orbital drawings of Si and a heavy post transition metal oxide (HPMO) between crystalline and amorphous structures is shown in Fig. (**5.17**) [82]. From the figure intuitively, the electron mobility decreases dramatically from c-silicon state to amorphous state, while medium mobility in crystalline HPMO is obtained even in the amorphous structure.

In a manner, the case of conduction band minimum in HPMO is quite similar to that in amorphous metal alloys regarding that metal orbitals predominantly constitute the electron path. The mobility of such amorphous state metal oxide can reach as large as 30-50 cm^2/Vs, such as indium tin oxide [83] and zinc tin oxide. Such high mobility in amor-phous materials is likely due to the conduction band mainly resulted from spherically symmetric, post-transition metal cation ns orbitals. The orbitals have large radii, resulting in overlap region between the adjacent orbitals and band dispersion. In addition, when compared to binary oxide semiconductors, such as ZnO, multicomponent oxide semiconductors have high possibility of amorphous state over a wide range of process parameters. Various amorphous oxide semiconductors have been studied and invest-igated, such as zinc indium oxide, zinc tin oxide, indium gallium oxide, indium gallium zinc oxide, and so on.

The TFTs using various stoichiometries of ZIO thin film deposited by rf sputtering as channel layer have been studied and investigated [84 - 86]. Even though the ZIO TFTs usually have high field effect mobility about 40 cm^2/Vs, but suffer from the issue that it is difficult to suppress the carrier concentration. So, the reported ZIO TFTs usually worked at depletion mode.

covalent semicon.   **ionic oxide semicon.**

crystal $M:(n-1)d^{10}ns^0 \ (n \geq 5)$

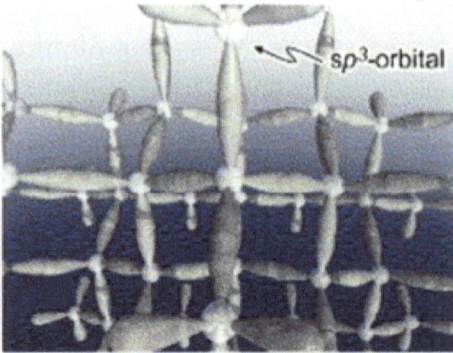
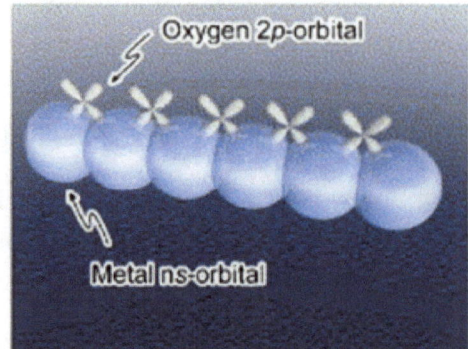

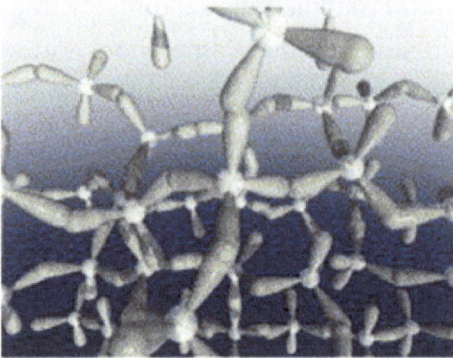
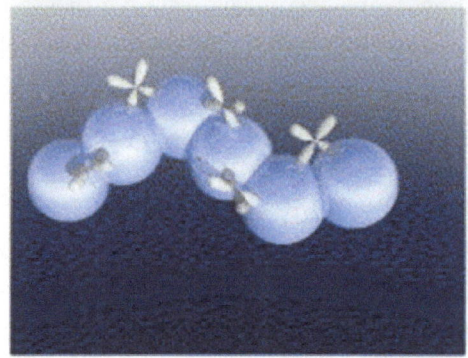

amorphous

Fig. (5.17). Schematic orbital drawing of electron pathway in covalent semiconductor and ionic oxide semiconductor [82].

In 2005, Dehuff *et al.* developed a high mobility TFTs using ZIO as active channel layer [86]. The ZIO thin film is deposited by rf magnetron sputtering technique at near room temperature. The ZIO TFTs was fabricated on glass substrate with ATO acting as the gate dielectric and transparent ITO as the gate and source/drain electrodes. Both TFTs working at enhancement mode and depletion mode are developed depending on anneal temperatures of 300 and 600 °C, respectively. The ZIO TFTs with depletion mode exhibit mobility of 45-55 cm^2/Vs, on/off drain current ratio of about 10^6, threshold voltages from -20 to -10 V, turn on voltage of about 3 V, and subthreshold swing of 0.8 V/decade. In contrast, the ZIO TFTs with enhancement mode show the highest mobility of 10-

30 cm^2/Vs, threshold voltage of 0-10 V, and turn on voltage of 1-2 V, on/off current ratios of 10^6, and subthreshold swing of 0.3 V/decade. Room temperature enhancement-mode ZIO TTFTs are also fabricated. Such devices exhibit highest field effect channel mobility of 8 cm^2/Vs and on/off current ratio of 10^4. The devices are highly transparent with about 85% optical transmission in the visible range.

Zinc Tin Oxide

In 2005, Chiang *et al.* demonstrated transparent bottom-gate structure TFTs using ZTO as active channel layer [87]. ZTO is a wide band gap material. The ZTO thin film was deposited by rf magnetron sputtering technique. The ZTO TFTs used an ATO layer acting as gate dielectric, and transparent ITO as gate and source/drain electrodes deposi-ted by rf magnetron sputtering technique. The ZTO TFTs exhibited a field-effect mobility of 5-15 cm^2/Vs and 20-50 cm^2/Vs with annealing temperature of 300 °C and 600 °C, respectively. The turn-on voltage of the devices is between -5 and 5 V and the on/off current ratio is about 10^7.

In 2005, staggered bottom-gate type ZTO-based TFTs were developed on flexible substrates by Jackson *et al.* [88]. The ZTO was deposited by sputtering technique. These TFTs utilize 375 nm thick SiON thin films as gate dielectric deposited by PECVD. The ZTO semiconductor was subjected to post-deposition annealing process at 250 °C for 10 minutes. The ZTO TFTs exhibit a field-effect mobility of 14 cm^2/Vs and a threshold voltage of -8 V. Great improvements were also achieved by optimizing some process condition and parameter. The threshold voltage can be shifted depending on the nitrogen content in the SiON layer and the oxygen vacancy in the ZTO active layer. The subthreshold swing is improved by reducing oxygen deficiency in the ZTO layer, because excess carriers degrade the performance, which is concluded from the measured capacitance-voltage characteristics.

Indium Gallium Oxide

In 2006, Presley *et al.* demonstrated Indium gallium oxide (IGO) TFTs and their application in transparent integrated circuits, such as inverters and ring oscillators [89]. A 100 nm thick SiO$_x$ layer deposited by PECVD was utilized as gate dielectric. Post-annealing process at 500 °C was carried out. The resulting IGO TFTs exhibited a field-effect mobility of 7 cm^2/Vs and turn-on voltage of 2 V. The maximum oscillation frequency of the ring oscillator circuit is only about 9.5 kHz due to the large parasitic capacitances.

Indium Gallium Zinc Oxide

In 2003, Nomura *et al.* proposed to utilize a complex $InGaO_3(ZnO)_5$ single-crystalline semiconductor layer as active channel layer in TFTs [90]. This layer was deposited by epitaxy method on an yttria-stabilized zirconia substrate and the proposed TFTs exhibited a high field effect mobility of 80 cm^2/Vs, turn-on voltage of about -0.5 V and on/off current ratio of 10^6. The annealing temperature is 1400 °C. However, it demonstrates that oxide semiconductor-based TFTs exhibit high electrical performance.

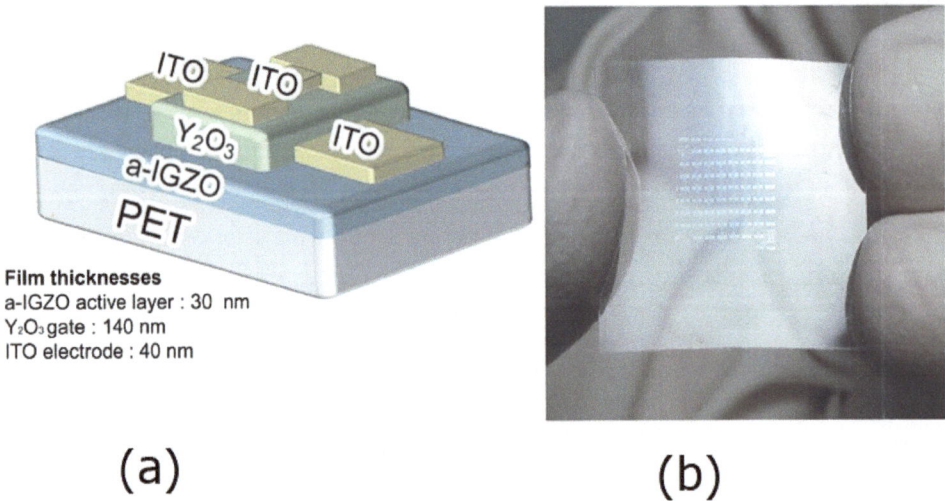

(a)   **(b)**

Fig. (5.18). Structure of the (**a**)-IGZO TFT fabricated on plastic sheet and (**b**) photograph of the flexible transparent TFTs sheet [91].

In 2004, Hosono *et al.* developed a novel amorphous oxide semiconductors based on indium gallium zinc oxide (IGZO) acted as active channel layer and the fabricated TFTs showed high electrical performance [91]. The a-IGZO is deposited by pulse laser deposition technique on polyethylene terephthalate at room temperature and shows hall effect mobility exceeding 10 cm^2/Vs, which is about 10 times larger than that of a-Si:H. The a-IGZO transparent TFT is a top-gate type structure with Y_2O_3 as gate dielectric and ITO as gate and source/drain electrodes, as shown in Fig. (**5.18**). The devices show saturation mobility of 6-9 cm^2/Vs and the device characteristics are stable during repe-titive bending of the flexible TFTs sheet.

This work attracted great attention in high resolution LCD and AMOLED flat panel display technology due to the high electron mobility, good uniformity, low process temperature, and suitability for large area application. The conduction band of IGZO thin film is related to the metal cation In 5 s orbital. The In 5 s orbital is spherical symmetry and isotropic, as shown in Fig. (**5.17**). So, the a-IGZO thin film still exhibits high electron mobility even in the amorphous state. In contrast, mobility of silicon semi-conductors drops significantly from the crystalline structure to the amorphous structure. In addition, the incorporation of Ga element is used as a carrier suppressor of the IZO thin film material and network stabilizer [82]. It is difficult to suppress the carrier concen-tration in the pure IZO thin film.

Since then, the optimization of metal cation composition [92, 93], gate dielectric materials [94, 95], and the device structures [96] are widely studied to improve the IGZO TFTs performance.

In 2007, Kim *et al.* proposed bottom-gate IGZO TFTs with SiO_x etch stopper. The devices exhibited a high field-effect mobility of 35.8 cm^2/Vs, an subthreshold swing value of 0.59 V/decade, and an on/off current ratio of 4.9×10^6 [96]. The a-IGZO active layer with a thickness of 50 nm was deposited by rf sputtering method on the SiO_2/glass substrate using a polycrystalline $In_2Ga_2ZnO_7$ target at room temperature. The inverted staggered bottom gate structure with 200 nm thick SiN_x as gate dielectric deposited by PECVD at 330 °C and MoW as gate and source/drain electrodes deposited by sputtering, as shown in Fig. (**5.19**). The adoption of the etch stopper SiO_x is necessary to prevent plasma-induced damage during the process and to achieve a-IGZO TFTs with high electrical performance.

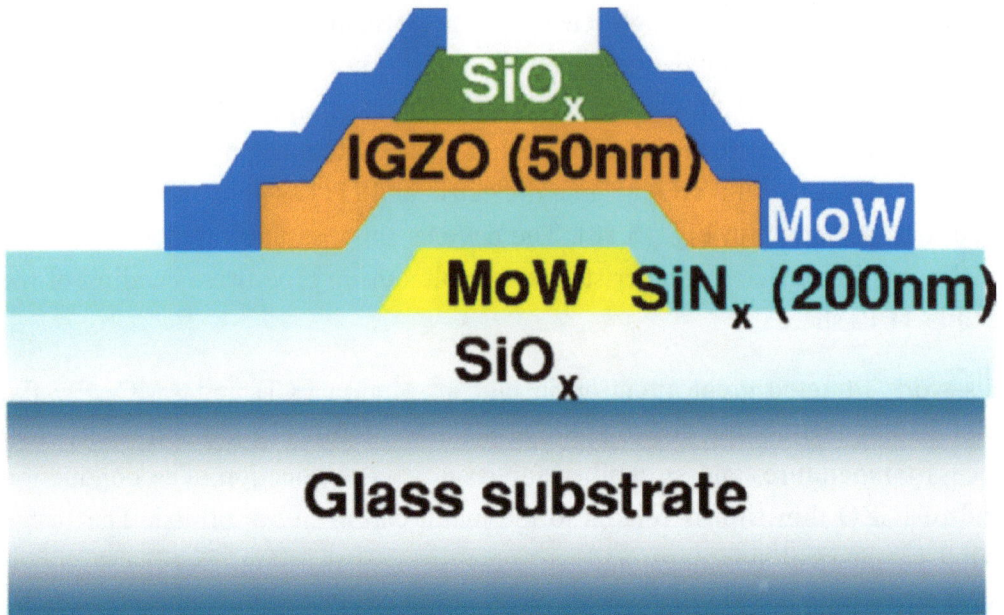

Fig. (5.19). Schematic cross section of the IGZO TFTs [96].

The bottom-gate structure of a-IGZO TFTs was widely studied and investigated in recent years. However, due to the high parasitic capacitance and poor scalability, it is unsuitable for the realization of complementary metal oxide semiconductor gates for digital and analog circuits. So, bottom-gate type a-IGZO TFTs could not be applied in the peripheral circuitry for the realization of SOP in the high end flat panel display. Several self-aligned top-gate structure was developed and demonstrated. The source/drain regions were heavily doped by hydrogen diffusion from silicon nitride passivation layer by PECVD [97], hydrogen or argon plasma treatment [98 - 101], or by metal reaction [102].

In 2008, Park *et al*. demonstrated self-aligned top gate IGZO TFTs with source/drain treated by Ar plasma, as shown in Fig. (**5.20a**). A 70 nm thick a-GIZO thin film was first sputtered by rf magnetron sputtering method at room temperature. The gate dielectric used SiO_2 layer deposited by PECVD at 150 °C and Mo was used as the gate, source/drain electrodes deposited by sputtering at room temperature. Low sheet resistance of the source/drain region in the a-GIZO TFT was obtained by Ar plasma treatment. The authors showed that the sheet resistance of the source/drain region was about 1 kΩ/□ after 60 seconds exposure time. The device exhibited good electrical performance including field effect mobility of 5 cm^2/Vs, subthreshold swing of 0.2 V/decade, threshold voltage of 0.2 V, and on/off current ratio of 10^7, as indicated in Fig. (**5.20b**).

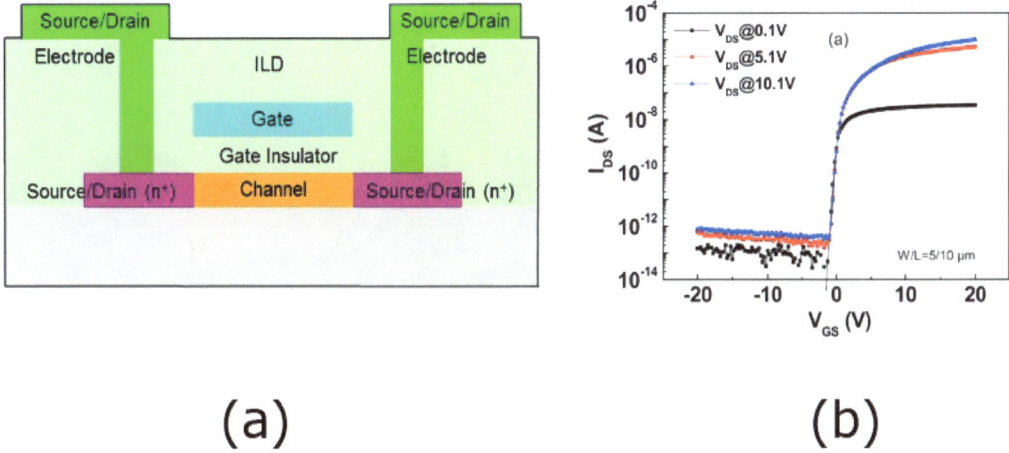

(a)   (b)

Fig. (5.20). (**a**) Schematic cross-sectional view of the self-aligned TFTs and (**b**) the transfer characteristic [99].

However, the TFT devices with argon or hydrogen treatment source/drain exhibit poor thermal stability. The hydrogen atom can rapidly diffuse in the a-IGZO materials at a temperature above 150 °C. Large amount of hydrogen diffuse out of the source/drain regions or into channel region, which will degrade the device performance.

Self-aligned top-gate a-IGZO TFTs with phosphorus or arsenic doped S/D regions has been developed [103, 104]. The S/D regions were heavily doped n-type using phosphorus or arsenic ion implantation technique.

The cross-sectional schematic of self-aligned top-gate a-IGZO TFT with phosphorus implanted source/drain is shown in Fig. (**5.21**). A 100 nm thick a-IGZO thin film was first sputtered on thermally oxidized silicon wafer by DC magnetron at room temperature. A 120 nm thick SiO_2 layer deposited by PECVD was used as the gate insulator. The gate electrode is a 230 nm thick ITO layer deposited by sputtering technique. The source/drain regions were self-aligned implanted with phosphorus element at a dose of 5×10^{15} /cm^2 and energy of 45 keV using gate electrode ITO pattern as a mask. Annealing process at 500 °C for 25 min in O_2 ambient was carried out to activate the implanted phosphorus dopant [103].

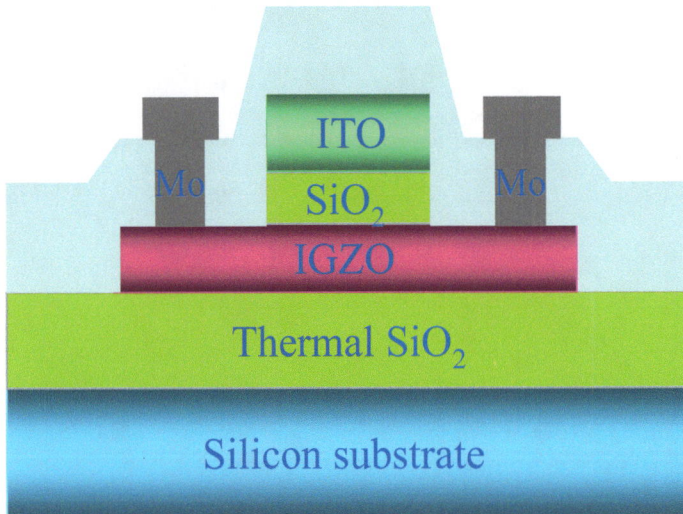

Fig. (5.21). Cross-section schematic of a-IGZO TFT with self-aligned top-gate structure.

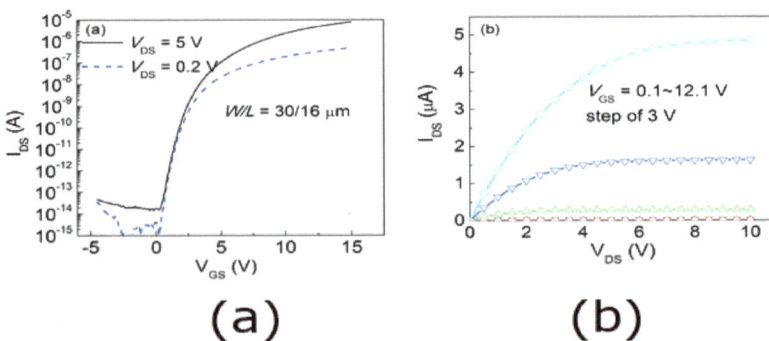

Fig. (5.22). (**a**) Transfer characteristic and (**b**) output characteristic of the a-IGZO TFTs [103].

The typical transfer and output characteristics of the self-aligned a-IGZO TFTs were shown in Fig. (**5.22**). They exhibit good transfer TFT characteristics at V_{DS} of 0.2 V such as field-effect mobility of 10 cm^2/Vs, threshold voltage of 5.5 V, subthreshold swing of 0.37 V/decade and on/off current ratio of 4×10^8. Low series resistance R_{SD} in S/D contacts were obtained from the clear linear regions in the output characteristic.

The hysteresis of *a*-IGZO TFT was examined, as shown in Fig. (**5.23**). Little shift of threshold voltage for the hysteresis loop was observed. It indicated that little electrons were trapped at or near the SiO$_2$/IGZO interface or within the a-IGZO channel layer.

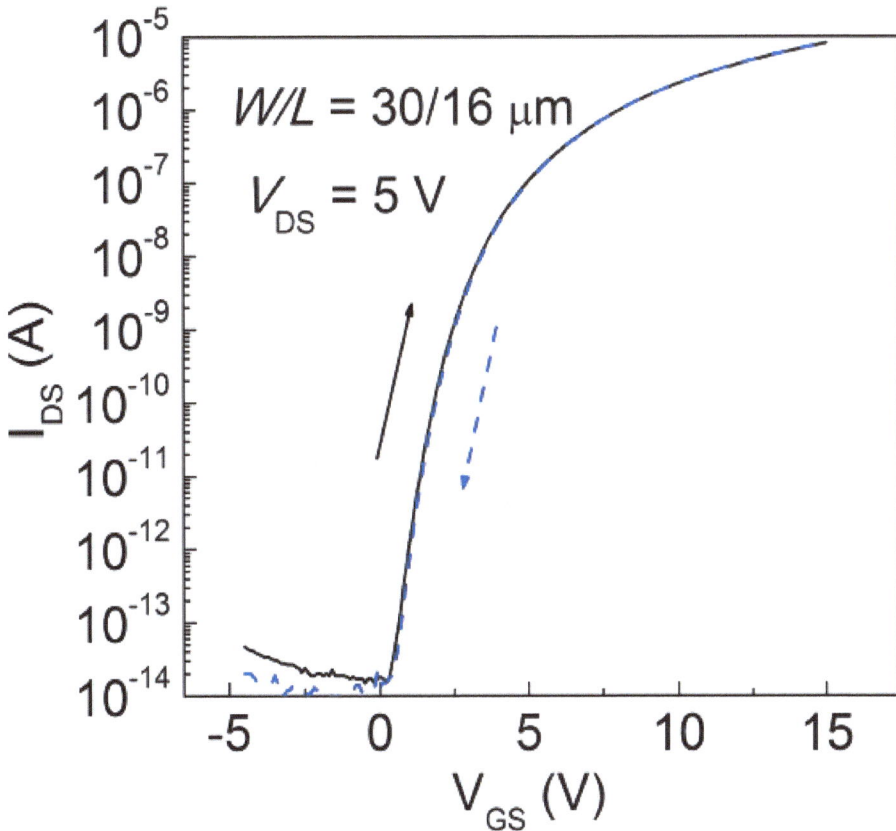

Fig. (5.23). Hysteresis characteristic of the a-IGZO TFTs.

Fig. (**5.24a**) shows the evolution of transfer characteristics for a-IGZO TFTs before and after heat-treatment at 200 °C for 0.5 hour and 1 hour. Very small degradation of subthreshold swing was observed. However, the performance of a-IGZO TFTs with S/D regions doped by argon or hydrogen plasma can easily degrade after heat-treatment at 200 °C, as shown Fig. (**5.24b**). The on-current I_{DS} decreased dramatically due to the diffusion of hydrogen out of the source/drain regions. If the annealing time increases, the hydrogen will diffuse into the whole channel, which will lead to large off-current and the device has difficulty to turn off [103].

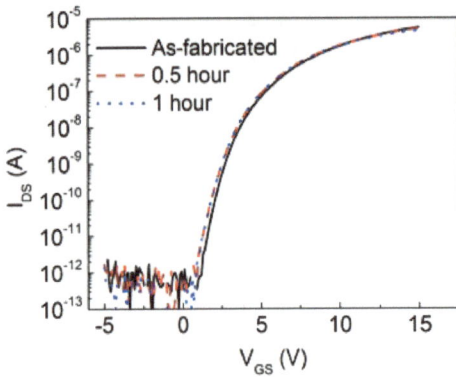

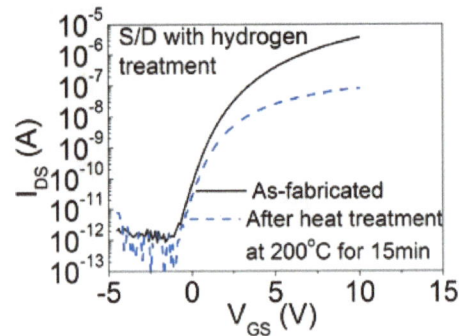

(a)   (b)

Fig. (5.24). Evolution of transfer characteristics of the a-IGZO TFTs with S/D regions (**a**) doped with phosphorus and (**b**) hydrogen under heat-treatment at 200°C.

GaN TFTs

GaN has emerged as one of the most promising compound semiconductor during the last few years. GaN-based high-electron-mobility transistors (HEMTs) have attracted significant attention in the area of high power, high speed, and high temperature transistors. High quality GaN thin film using inexpensive substrate under low deposition temperature is desired. Amorphous and polycrystalline GaN thin films were deposited using magnetron sputtering technique [105 - 108] or pulsed laser deposition technique [109]. For the mass production in display industry, DC magnetron sputtering technique has the advantages of high deposition rate, large area, good uniformity and low cost.

In 1998, Bottom-gate type and top-gate type GaN TFTs proposed by Kobayashi *et al*. exhibited poor electrical performance, due to localized gap states in GaN thin film and large series resistance at the interface between GaN film and Al electrode [110]. The authors reported that the nano-crystalline GaN was deposited by a reactive sputtering method in a mixed N_2 and Ar atmosphere using metal Ga as sputtering target. Fig. (**5.25**) shows the XRD pattern of the nc-GaN thin film reported by Kobayashi *et al*. The diffraction peaks were obtained at around 32.5°, 36.9° and 57.9° corresponding to (100), (101) and (110) of hexagonal crystalline GaN.

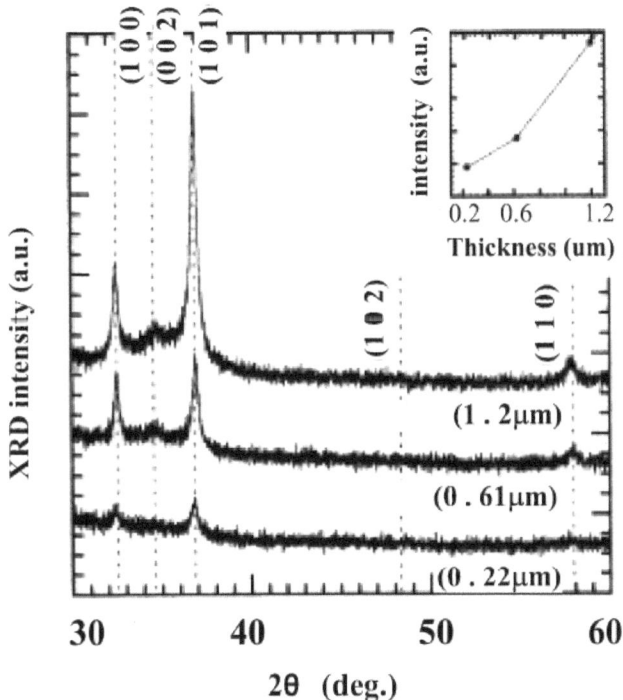

Fig. (5.25). The XRD pattern of nc-GaN films reported by Kobayashi *et al*. [111].

The structure of a bottom-gate TFT using GaN as active channel layer is shown in Fig. (**5.26**). A n-type crystal silicon wafer covered with SiO_2 acted as gate dielectric was used as gate electrode. The source and drain electrodes were Al thin films deposited by evaporation technique at room temperature. Fig. (**5.27**) shows the transfer characteristics of the bottom-gate type GaN TFT. Kobayashi *et al.* proposed two annealing process. The nc-GaN thin films were annealed at 800 °C for 1 hour in vacuum (Process A) to reduce the density of localized states. The device was annealed at 150 °C (Process B) after the deposition of the drain and source electrodes. In Fig. (**5.27**), TFT-I denoted the devices without any annealing process. TFT-II denoted the devices with annealing Process A. TFT-III denoted the devices with annealing Process A and B. All of the devices showed poor performance, including the TFT-III devices with low mobility (6×10^{-2} cm^2/Vs) and low on/off current ratio (3×10^3), due to localized gap states in GaN thin film and large series resistance at the interface between GaN film and Al electrode.

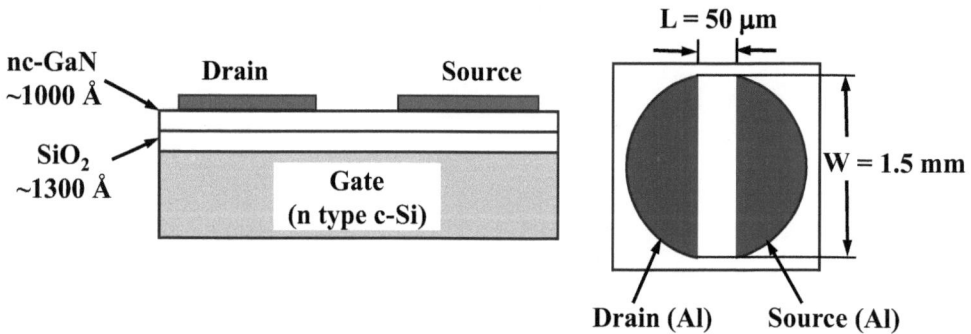

Fig. (5.26). Cross-sectional and top view of GaN TFT reported by Kobayashi *et al.* [110].

In 2012, top-gate GaN TFTs with implanted silicon source/drain regions were also deve-loped [112]. With low source/drain series resistance, the GaN TFT showed much better performance. GaN thin film was prepared by reactive DC magnetron sputtering at room temperature. The cross-sectional schematic of the GaN TFTs is shown in Fig. (**5.28**). A 150 nm thick GaN thin film was sputtered on thermally oxidized silicon wafer substrate. Source and drain regions were selectively implanted with silicon at dose of 3×10^{15} /cm^2 and energy of 190 keV through 25 nm PECVD oxide. Annealing process at 1100°C for 5 min in N_2 ambient was carried out for the activation of implanted silicon dopant. A SiO_2 layer deposited by PECVD acted as gate insulator. The ohmic contacts on source and drain region were formed using Ti/Al double metal layers deposited by sputtering.

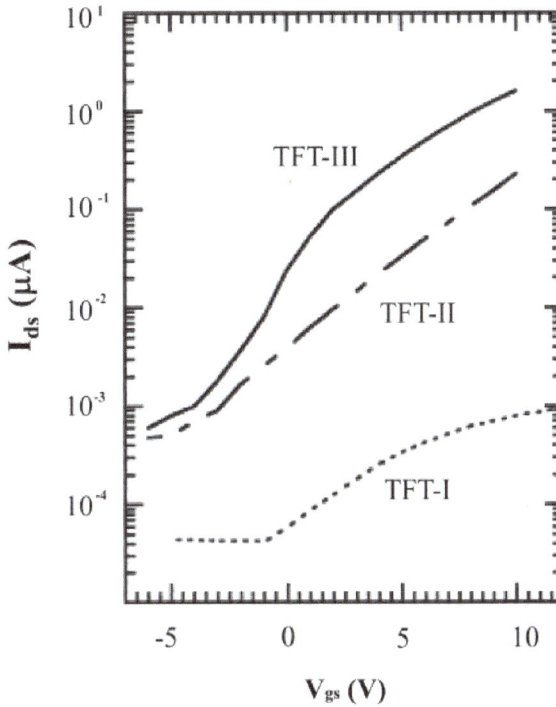

Fig. (5.27). Transfer characteristics of GaN TFTs [110].

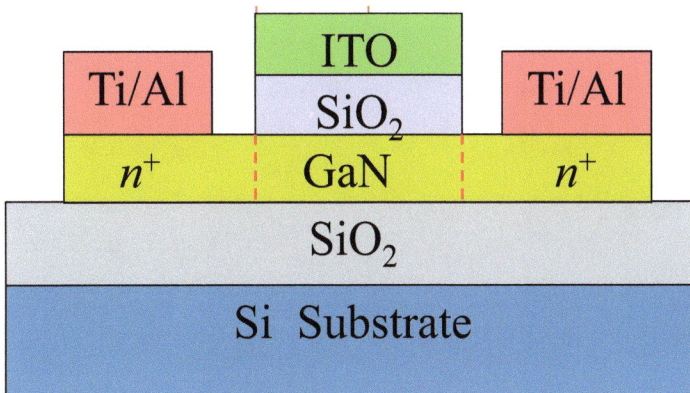

Fig. (5.28). Cross-sectional schematic of the GaN TFTs [112].

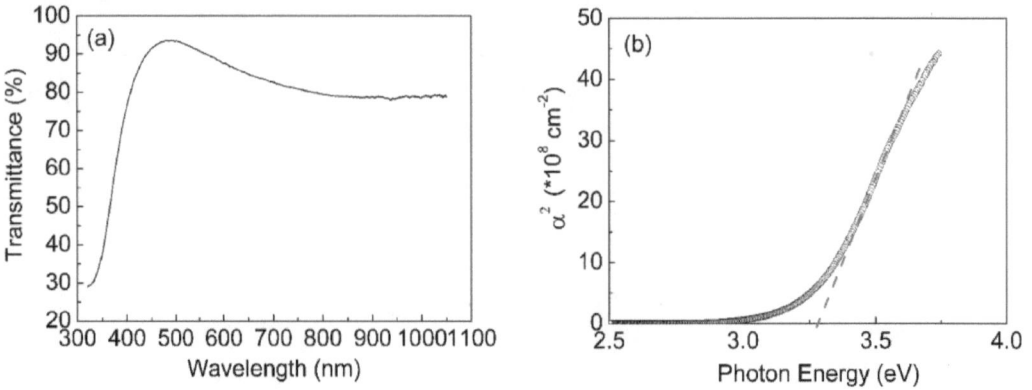

Fig. (5.29). (**a**) Optical transmittance spectrum for the GaN thin film with a thickness of 150 nm and (**b**) the square of absorption coefficient as a function of photon energy for the GaN films [112].

The transmittance curves of the 150 nm thick GaN thin film is shown in Fig. (**5.29a**), where it can be seen that the film show 80-93% optical transmission in the visible range. This high transmittance is important for transparent electronics applications. The optical band gap has been derived to be about 3.3 eV, as shown in Fig. (**5.29b**).

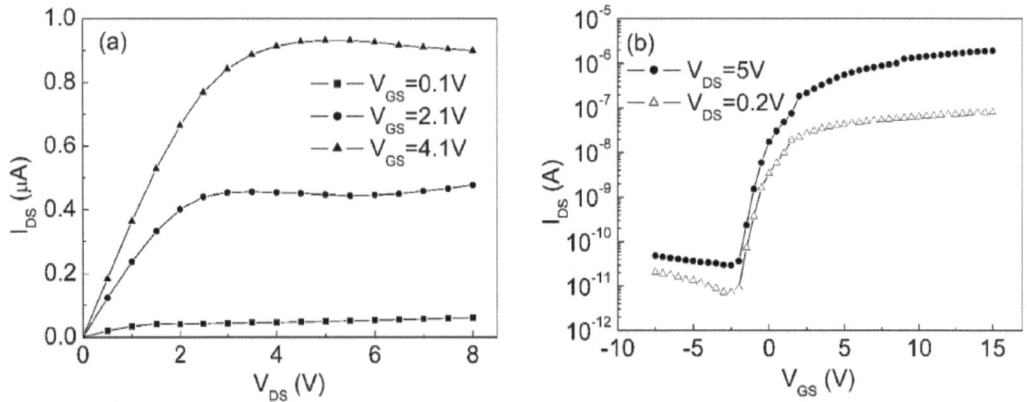

Fig. (5.30). (**a**) Output characteristics and (**b**) transfer characteristics of the GaN TFTs (W/L = 10/2 ,m).

The typical output characteristics of the n-type GaN TFTs with a channel width to length ratio of 10 μm/2 μm are shown in Fig. (**5.30a**). From the clear linear regions in the output characteristic, low series resistance in source/drain contacts were obtained due to the heavily doped GaN in source/drain region and the activation of the silicon dopant.

The transfer characteristics with V_{DS} =0.2 V and 5 V for the GaN TFTs are shown in Fig. (**5.30b**). They exhibit a field effect mobility of 1 cm²/Vs, a threshold voltage of -0.4 V, a subthreshold swing of 0.8 V/decade, and an on/off current ratio of 10^5.

MoS₂ TFTs

Recently, due to its interesting electrical and optical properties, single-layer MoS₂ has attracted lots of investigation and studies. While bulk MoS₂ is an n-type material with an indirect bandgap (~1.3 eV), single-layer MoS₂ has a direct bandgap of ~1.8 eV [113]. Field-effect transistors using single-layer MoS₂ showed high on/off current ratios (~10^8) and steep subthreshold behavior (~70mV/decade) [114]. The electron mobility of single-layer MoS₂ FETs varied from ~1 cm²/Vs to ~200 cm²/Vs depending on the dielectric. MoS₂ can be an attractive alternative for TFTs active layer application for high-resolution AMOLED displays. In 2012, S. Kim *et al.* [115] developed multilayer MoS₂ TFTs with high mobility (>100 cm²/Vs), near ideal subthreshold swing (~70 mV/decade). The TFTs based on multilayer MoS₂ were fabricated, as shown in Fig. (**5.31**). Amorphous 50 nm thick Al₂O₃ dielectric layer was grown on a heavily doped p-type Si wafer by ALD process. Multilayer MoS₂ flakes were mechanically exfoliated from bulk MoS₂ crystals and transferred on ALD-Al₂O₃-covered Si substrates. Ti/Au double metal layer deposited by evaporation method were used as the source/drain electrodes. Post-annealing process at 200 °C in vacuum for 2 hours at Ar/H₂ atmosphere was performed to improve the contacts.

Fig. (**5.32a**) shows the transfer characteristic of the n-type MoS₂ TFTs with room-temperature mobility of larger than 100 cm²/Vs and very low *SS* of 70 mV/decade. Fig. (**5.32b**) shows the output characteristic of the TFTs. The drain current show pinch-off and strong saturation behavior at high drain to source biases for all gate voltages.

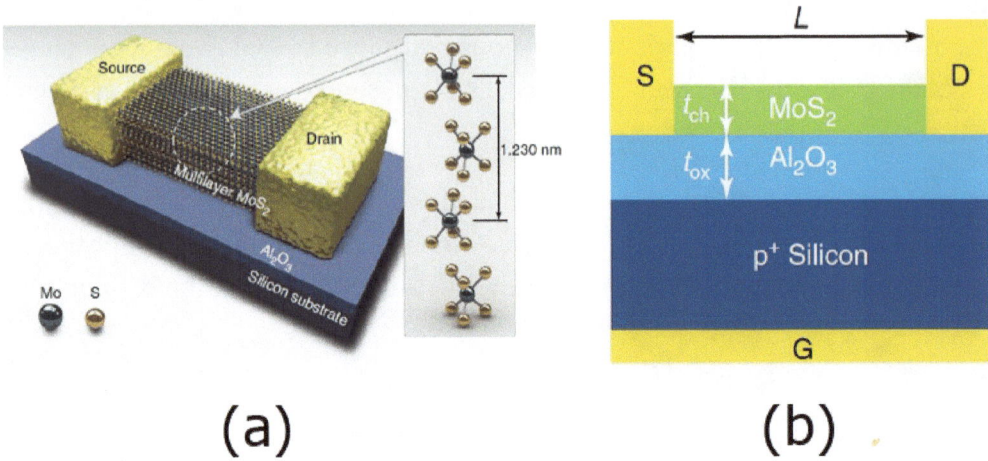

Fig. (5.31). (**a**) Schematic perspective view and (**b**) the device geometry of a MoS2 TFT with a multilayer MoS2 crystal [115].

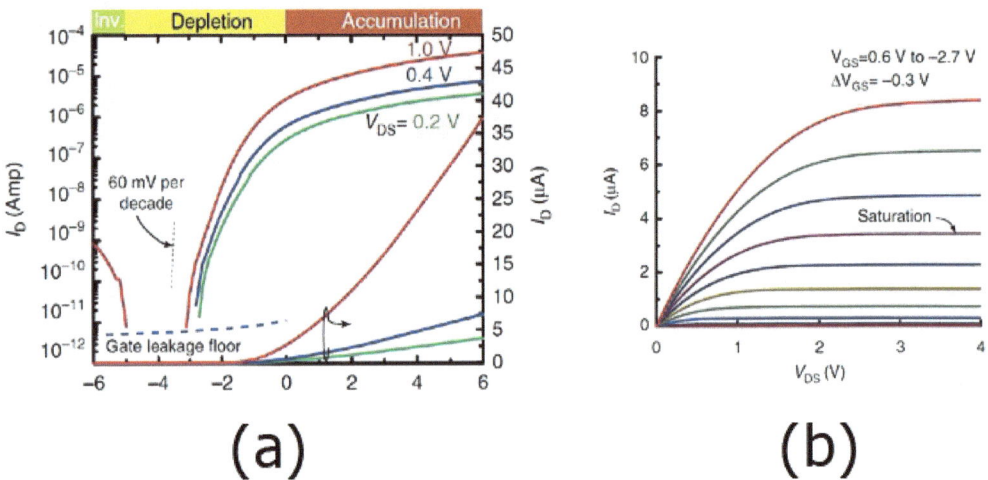

Fig. (5.32). (**a**) the transfer and (**b**) output characteristic of the multiple-layer MoS2 TFTs [115].

However, the active layer MoS_2 is formed using exfoliation method, which is not suitable for large area application. The large area deposition method for MoS_2 thin film is still needed for investigation.

SUMMARY

TFT device is an essential part for realizing high resolution, large size, and high quality AMOLED display. Several main TFTs technologies are introduced in this chapter. ELA LTPS TFTs based backplanes have been commercialized and widely applied in the high-end smart phone market. However, laser annealing technology suffers from the high cost of equipment and limits the size of the substrates. The development of new type TFT technology with high electrical performance, good uniformity, high stability and reliability are necessary for high quality AMOLED display.

CONFLICT OF INTEREST

The authors confirm that this chapter contents have no conflict of interest.

ACKNOWLEDGEMENTS

Declared none.

REFERENCES

[1]　Y. Chang, M.K. Wei, C.M. Kou, S.J. Shieh, J.H. Lee, and C.C. Chen, "Manufacture of passive matrix OLED-organic light emitting display", *SID Symposium Digest of Technical Papers,* pp. 1040-1043, 2001.

[2]　Z. Meng, H. Chen, C. Qiu, H.S. Kwok, and M. Wong, "Active-matrix organic light-emitting diode display implemented using metal-induced unilaterally crystallized polycrystalline silicon thin-film transistors", *SID Symposium Digest of Technical Papers,* pp. 380-383, 2001.
[http://dx.doi.org/10.1889/1.1831875]

[3]　J. Yamashita, K. Uchino, T. Yamamoto, T. Sasaoka, and T. Urabe, "New driving method with current subtraction pixel circuit for AM-OLED displays", *SID Symposium Digest of Technical Papers,* pp. 1452-1455, 2005.
[http://dx.doi.org/10.1889/1.2036281]

[4]　Z. Meng, *"Metal-induced unilaterally crystallized polycrystalline silicon thin-film transistor technology and application to flat-panel displays"*, Thesis (Ph.D.)-HK University of Science and Technology, 2002.

[5]　T. Sasaoka, M. Sekiya, and A. Yumoto, "A 13.0 inch AMOLED display with top emitting structure and adaptive current mode programmed pixel circuit (TAC)", *SID Symposium Digest of Technical Papers,* pp. 384-387, 2001.

[6]　P.K. Weimer, "The TFT - a new thin-film transistor", *Proceeding of the Institute of Radio Engineers,* pp. 1462-1469, 1963.

[http://dx.doi.org/10.1109/JRPROC.1962.288190]

[7]   F.V. Shallcross, "Cadmium selenide thin-film transistors", In: *Proceedings of the IEEE.* 1963, p. 851.

[8]   P.K. Weimer, "A p-type tellurium thin-film transistor", In: *Proceedings of the IEEE.* 1964, pp. 608-609.

[9]   S.R. Hofstein, and F.P. Heiman, "The silicon insulated-gate field-effect transistor", *Proceeding of the IEEE.*, pp. 1190-1202, 1963.

[10]   T.P. Brody, A.A. Juris, and G.D. Dixon, "A 6×6 inch 20 lines-per-inch liquid crystal display panel", In: *IEEE Trans. Electron. Dev,* vol. 20. 1973, no. 11, pp. 995-1001.
        [http://dx.doi.org/10.1109/T-ED.1973.17780]

[11]   T.P. Brody, "The thin-film transistor - a late flowering bloom", *IEEE Trans. Electron. Dev.,* vol. 31, no. 11, pp. 1614-1628, 1984.
        [http://dx.doi.org/10.1109/T-ED.1984.21762]

[12]   M.J. Lee, S.W. Wright, and C.P. Judge, "Electrical and structural properties of cadmium selenide thin-film transistors", *Solid-State Electron.,* vol. 23, no. 6, pp. 671-679, 1980.
        [http://dx.doi.org/10.1016/0038-1101(80)90053-2]

[13]   W.E. Spear, P.G. Comber, S. Kinmond, and M.H. Brodsky, "Amorphous silicon p-n junction", *Appl. Phys. Lett.,* vol. 28, pp. 105-107, 1976.
        [http://dx.doi.org/10.1063/1.88658]

[14]   P.G. Lecomber, W.E. Spear, and A. Ghaith, "Amorphous-silicon field-effect device and possible application", *Electron. Lett.,* vol. 15, no. 6, pp. 179-181, 1979.
        [http://dx.doi.org/10.1049/el:19790126]

[15]   D. Murley, N. Young, M. Trainor, and D. McCulloch, "An investigation of laser annealed and metal-induced crystallized polycrystalline silicon thin-film transistors", *IEEE Trans. Electron. Dev.,* vol. 48, no. 6, pp. 1145-1151, 2001.
        [http://dx.doi.org/10.1109/16.925240]

[16]   M.K. Hatalis, and D.W. Greve, "High performance thin-film transistors in low-temperature crystallized LPCVD amorphous silicon films", *IEEE Electron Device Lett.,* vol. 8, no. 8, pp. 361-364, 1987.
        [http://dx.doi.org/10.1109/EDL.1987.26660]

[17]   C. Spinella, S. Lombardo, and F. Priolo, "Crystal grain nucleation in amorphous silicon", *J. Appl. Phys.,* vol. 84, pp. 5383-5414, 1998.
        [http://dx.doi.org/10.1063/1.368873]

[18]   N. Yamauchi, and R. Reif, "Polycrystalline silicon thin-films processed with silicon ion-implantation and subsequent solid-phase crystallization - theory, experiments, and thin-film-transistor applications", *J. Appl. Phys.,* vol. 75, pp. 3235-3257, 1994.
        [http://dx.doi.org/10.1063/1.356131]

[19]   T. Noguchi, H. Hayashi, and T. Ohshima, "Low-temperature polysilicon super-thin-film transistor (LSFT)", *Jpn. J. Appl. Phys.,* vol. 25, no. 2, pp. L121-L123, 1986.
        [http://dx.doi.org/10.1143/JJAP.25.L121]

[20] C. Chang, "Enhanced performance and reliability for solid-phase crystallized poly-Si TFTs with argon ion implantation", *J. Electrochem. Soc.,* vol. 154, pp. J375-J378, 2007.
[http://dx.doi.org/10.1149/1.2778858]

[21] M. Ryu, S. Hwang, T. Kim, K. Kim, and S. Min, "The effect of surface nucleation on the evolution of crystalline microstructure during solid phase crystallization of amorphous Si films on SiO2", *Appl. Phys. Lett.,* vol. 71, pp. 3063-3065, 1997.
[http://dx.doi.org/10.1063/1.119437]

[22] Y. Kuo, and P. Kozlowski, "Polycrystalline silicon formation by pulsed rapid thermal annealing of amorphous silicon", *Appl. Phys. Lett.,* vol. 69, pp. 1092-1094, 1996.
[http://dx.doi.org/10.1063/1.117068]

[23] M. Bonnel, N. Duhamel, M. Guendouz, L. Haji, B. Loisel, and P. Ruault, "Poly-Si thin-film transistors fabricated with rapid thermal annealed silicon films", *Jpn. J. Appl. Phys.,* vol. 30, no. 11, pp. L1924-L1926, 1991.
[http://dx.doi.org/10.1143/JJAP.30.L1924]

[24] V. Subramanian, P. Dankoski, L. Degertekin, B. KhuriYakub, and K. Saraswat, "Controlled two-step solid-phase crystallization for high-performance polysilicon TFT's", *IEEE Electron Device Lett.,* vol. 18, no. 8, pp. 378-381, 1997.
[http://dx.doi.org/10.1109/55.605445]

[25] H. Seo, C. Kim, and I. Kang, "Alternating magnetic field-assisted crystallization of Si films without metal catalyst", *J. Cryst. Growth,* vol. 310, no. 12, pp. 5317-5320, 2008.
[http://dx.doi.org/10.1016/j.jcrysgro.2008.09.182]

[26] H. Seo, D.H. Nam, and N.B. Choi, "Low cost and uniform solid phase crystallization without metal catalyst employing alternating magnetic field for AM-OLED", In: *12th International Display Workshops/Asia Display*, 2005, pp. 1127-1132.

[27] S. Hong, B. Kim, and Y. Ha, "LTPS technology for improving the uniformity of AMOLEDs", *SID Symposium Digest of Technical Papers*, pp. 1366-1369, 2007.

[28] S.H. Jung, H.K. Lee, C.Y. Kim, S.Y. Yoon, C.D. Kim, and I.D. Kang, "15-inch AMOLED display with SPC TFTs and a symmetric driving method", *SID Symposium Digest of Technical Papers*, pp. 101-104, 2008.
[http://dx.doi.org/10.1889/1.3069305]

[29] J. Lee, C. Lee, and D. Choi, "Influences of various metal elements on field aided lateral crystal-lization of amorphous silicon films", *Jpn. J. Appl. Phys.,* vol. 40, no. 11, pp. 6177-6181, 2001.
[http://dx.doi.org/10.1143/JJAP.40.6177]

[30] Z. Jin, G. Bhat, M. Yeung, H.S. Kwok, and M. Wong, "Nickel induced crystallization of amorphous silicon thin films", *J. Appl. Phys.,* vol. 84, pp. 194-200, 1998.
[http://dx.doi.org/10.1063/1.368016]

[31] S. Yoon, S. Park, K. Kim, J. Jang, and C. Kim, "Structural and electrical properties of polycrystalline silicon produced by low-temperature Ni silicide mediated crystallization of the amorphous phase", *J. Appl. Phys.,* vol. 87, pp. 609-611, 2000.
[http://dx.doi.org/10.1063/1.371906]

[32]   S. Zhao, Z. Meng, C. Wu, S. Xiong, M. Wong, and H.S. Kwok, "Solution-based metal induced crystallized polycrystalline silicon films and thin-film transistors", *J. Mater. Sci. Mater. Electron.,* vol. 18, no. 10, pp. S117-S121, 2007.
[http://dx.doi.org/10.1007/s10854-007-9157-0]

[33]   J. Choi, D. Kim, S. Kim, S. Park, and J. Jang, "Polycrystalline silicon prepared by metal induced crystallization", *Thin Solid Films,* vol. 440, no. 9, pp. 1-4, 2003.
[http://dx.doi.org/10.1016/S0040-6090(03)00821-6]

[34]   J. Jang, J. Oh, S. Kim, Y. Choi, S. Yoon, and C. Kim, "Electric-field-enhanced crystallization of amorphous silicon", *Nature,* vol. 395, no. 10, pp. 481-483, 1998.
[http://dx.doi.org/10.1038/26711]

[35]   J. Jang, S. Park, K. Kim, B. Cho, W. Kwak, and S. Yoon, "Polycrystalline silicon produced by Ni-silicide mediated crystallization of amorphous silicon in an electric field", *J. Appl. Phys.,* vol. 88, pp. 3099-3101, 2000.
[http://dx.doi.org/10.1063/1.1286064]

[36]   S. Yoon, S. Park, K. Kim, and J. Jang, "Metal-induced crystallization of amorphous silicon", *Thin Solid Films,* vol. 383, no. 2, pp. 34-38, 2001.
[http://dx.doi.org/10.1016/S0040-6090(00)01790-9]

[37]   H. Kim, B. Kim, J. Bae, K. Hwang, H. Seo, and C. Kim, "Kinetics of electric-field-enhanced crystallization of amorphous silicon in contact with Ni catalyst", *Appl. Phys. Lett.,* vol. 81, pp. 5180-5182, 2002.
[http://dx.doi.org/10.1063/1.1532533]

[38]   J. Choi, S. Kim, J. Cheon, S. Park, Y. Son, and J. Jang, "Kinetics of Ni-mediated crystallization of a-Si through a SiNx cap layer", *J. Electrochem. Soc.,* vol. 151, pp. G448-G451, 2004.
[http://dx.doi.org/10.1149/1.1752936]

[39]   J. Choi, D. Kim, B. Choo, W. Sohn, and J. Jang, "Metal induced lateral crystallization of amorphous silicon through a silicon nitride cap layer", *Electrochem. Solid-State Lett.,* vol. 6, no. 1, pp. G16-G18, 2003.
[http://dx.doi.org/10.1149/1.1527411]

[40]   W. Sohn, J. Choi, K. Kim, J. Oh, S. Kim, and J. Jang, "Crystalline orientation of polycrystalline silicon with disklike grains produced by silicide-mediated crystallization of amorphous silicon", *J. Appl. Phys.,* vol. 94, pp. 4326-4331, 2003.
[http://dx.doi.org/10.1063/1.1604958]

[41]   J. Kim, J. Choi, S. Kim, and J. Jang, "Stable polycrystalline silicon TFT with MICC", *IEEE Electron Device Lett.,* vol. 25, no. 4, pp. 182-184, 2004.
[http://dx.doi.org/10.1109/LED.2004.824844]

[42]   S. Lee, T. Ihn, and S. Joo, "Fabrication of high-mobility p-channel poly-Si thin film transistors by self-aligned metal-induced lateral crystallization", *IEEE Electron Device Lett.,* vol. 17, no. 8, pp. 407-409, 1996.
[http://dx.doi.org/10.1109/55.511590]

[43]   S. Lee, and S. Joo, "Low temperature poly-Si thin-film transistor fabrication by metal-induced lateral

crystallization", *IEEE Electron Device Lett.,* vol. 17, no. 4, pp. 160-162, 1996.
[http://dx.doi.org/10.1109/55.485160]

[44]   S. Jun, Y. Yang, J. Lee, and D. Choi, "Electrical characteristics of thin-film transistors using field-aided lateral crystallization", *Appl. Phys. Lett.,* vol. 75, pp. 2235-2237, 1999.
[http://dx.doi.org/10.1063/1.124975]

[45]   M. Wong, Z. Jin, G. Bhat, P. Wong, and H.S. Kwok, "Characterization of the MIC/MILC interface and its effects on the performance of MILC thin-film transistors", *IEEE Trans. Electron. Dev.,* vol. 47, no. 5, pp. 1061-1067, 2000.
[http://dx.doi.org/10.1109/16.841241]

[46]   Z. Meng, H. Chen, C. Qiu, L. Wang, and H.S. Kwok, "Application of metal-induced unilaterally crystallized polycrystalline silicon thin-film transistor technology to active-matrix organic light-emitting diode displays", *IEEE Electron Devices Meeting,* pp. 611-614, 2000.

[47]   Z. Meng, and M. Wang, "M. Wong M, "High performance low temperature metal-induced unilaterally crystallized polycrystalline silicon thin film transistors for system-on-panel applications", *IEEE Trans. Electron. Dev.,* vol. 47, no. 2, pp. 404-409, 2000.
[http://dx.doi.org/10.1109/16.822287]

[48]   Z. Meng, and M. Wong, "Active-matrix organic light-emitting diode displays realized using metal-induced unilaterally crystallized polycrystalline silicon thin-film transistors", *IEEE Trans. Electron. Dev.,* vol. 49, no. 6, pp. 991-996, 2002.
[http://dx.doi.org/10.1109/TED.2002.1003718]

[49]   G. Fortunato, "Polycrystalline silicon thin-film transistors: A continuous evolving technology", *Thin Solid Films,* vol. 296, no. 3, pp. 82-90, 1997.
[http://dx.doi.org/10.1016/S0040-6090(96)09378-9]

[50]   T. Noguchi, A. Tang, J. Tsai, and R. Reif, "Comparison of effects between large-area-beam ELA and SPC on TFT characteristics", *IEEE Trans. Electron. Dev.,* vol. 43, no. 9, pp. 1454-1458, 1996.
[http://dx.doi.org/10.1109/16.535332]

[51]   A. Marmorstein, A. Voutsas, and R. Solanki, "A systematic study and optimization of parameters affecting grain size and surface roughness in excimer laser annealed polysilicon thin films", *J. Appl. Phys.,* vol. 82, pp. 4303-4309, 1997.
[http://dx.doi.org/10.1063/1.366238]

[52]   H. Kuriyama, S. Kiyama, and S. Noguchi, "Enlargement of poly-Si film grain-size by excimer laser annealing and its application to high-performance poly-Si thin-film transistor", *Jpn. J. Appl. Phys.,* vol. 30, no. 12, pp. 3700-3703, 1991.
[http://dx.doi.org/10.1143/JJAP.30.3700]

[53]   P. Mei, J. Boyce, and M. Hack, "Laser dehydrogenation crystallization of plasma-enhanced chemical-vapor-deposited amorphous-silicon for hybrid thin-film transistors", *Appl. Phys. Lett.,* vol. 64, pp. 1132-1134, 1994.
[http://dx.doi.org/10.1063/1.110829]

[54]   C. Angelis, and C. Dimitriadis, "Effect of excimer laser annealing on the structural and electrical properties of polycrystalline silicon thin-film transistors", *J. Appl. Phys.,* vol. 86, pp. 4600-4606, 1999.
[http://dx.doi.org/10.1063/1.371409]

[55]   A. Voutsas, A. Limanov, and J. Im, "Effect of process parameters on the structural characteristics of laterally grown, laser-annealed polycrystalline silicon films", *J. Appl. Phys.*, vol. 94, pp. 7445-7452, 2003.
[http://dx.doi.org/10.1063/1.1627462]

[56]   J. Im, M. Crowder, and R. Sposili R, "Controlled super-lateral growth of Si films for microstructural manipulation and optimization", *Physica Status Solidi A Appl. Res.*, vol. 166, pp. 603-617, 1998.

[57]   H. Song, and J. Im, "Excimer-laser crystallization of patterned Si films at high temperatures via artificially controlled super-lateral growth", *Advanced Laser Processing of Materials - Fundamentals Applications*, vol. 397, pp. 459-464, 1996.

[58]   C. Oh, and M. Matsumura, "A proposed single grain-boundary thin-film transistor", *IEEE Electron Device Lett.*, vol. 22, no. 1, pp. 20-22, 2001.
[http://dx.doi.org/10.1109/55.892431]

[59]   A. Hara, M. Takei, and F. Takeuchi, "High performance low temperature polycrystalline silicon thin film transistors on non-alkaline glass produced using diode pumped solid state continuous wave laser lateral crystallization", *Jpn. J. Appl. Phys.*, vol. 43, no. 4, pp. 1269-1276, 2004.
[http://dx.doi.org/10.1143/JJAP.43.1269]

[60]   M. Tai, M. Hatano, S. Yamaguchi, T. Noda, S. Park, and T. Shiba, "Performance of poly-Si TFTs fabricated by SELAX", *IEEE Trans. Electron. Dev.*, vol. 51, no. 6, pp. 934-939, 2004.
[http://dx.doi.org/10.1109/TED.2004.828167]

[61]   J. Boyce, P. Mei, R. Fulks, and J. Ho, "Laser processing of polysilicon thin-film transistors: grain growth and device fabrication", *Phys. Status Solidi, A Appl. Res.*, vol. 166, pp. 729-741, 1998.
[http://dx.doi.org/10.1002/(SICI)1521-396X(199804)166:2<729::AID-PSSA729>3.0.CO;2-1]

[62]   S. Zhao, Z. Meng, and X. Li, "Metal induced continuous zonal domain polycrystalline silicon and thin film transistors", *SID Symposium Digest of Technical Papers*, pp. 233-236, 2007.
[http://dx.doi.org/10.1889/1.2785272]

[63]   W. Zhou, Z. Meng, and S. Zhao S, ""Bridged-grain solid-phase-crystallized polycrystalline-silicon thin-film transistors"", *IEEE Electron Devices Lett*, vol. 33, no. 10, pp. 1414-1416, 2012.

[64]   R. Proano, R. Misage, and D. Ast, "Development and electrical properties of undoped polycrystalline silicon thin-film transistors", *IEEE Trans. Electron. Dev.*, vol. 36, no. 9, pp. 1915-1922, 1989.
[http://dx.doi.org/10.1109/16.34270]

[65]   C. Kim, K. Sohn, and J. Jang, "Temperature dependent leakage currents in polycrystalline silicon thin film transistors", *J. Appl. Phys.*, vol. 81, pp. 8084-8090, 1997.
[http://dx.doi.org/10.1063/1.365416]

[66]   J.F. Wager, and R. Hoffman, "Thin, fast, and flexible", *IEEE Spectr.*, vol. 48, no. 5, pp. 42-56, 2011.
[http://dx.doi.org/10.1109/MSPEC.2011.5753244]

[67]   R. Martins, I. Ferreira, and E. Fortunato, "Electronics with and on paper", *Phys. Status Solidi Rapid Res. Lett.*, vol. 5, no. 9, pp. 332-335, 2011.
[http://dx.doi.org/10.1002/pssr.201105247]

[68]   U. Ozgur, Y.I. Alivov, and C. Liu, "A comprehensive review of ZnO materials and devices", *J. Appl. Phys.*, vol. 98, p. 041301, 2005.

[http://dx.doi.org/10.1063/1.1992666]

[69]   L.J. Meng, and M.P. Dossantos, "Direct-current reactive magnetron sputtered zinc-oxide thin-films the effect of the sputtering pressure", *Thin Solid Films,* vol. 250, no. 10, pp. 26-32, 1994.
[http://dx.doi.org/10.1016/0040-6090(94)90159-7]

[70]   T. Minami, H. Nanto, and S. Takata, "Highly conductive and transparent aluminum doped zinc-oxide thin-films prepared by rf magnetron sputtering", *Jpn. J. Appl. Phys.,* vol. 23, pp. L280-L282, 1984.
[http://dx.doi.org/10.1143/JJAP.23.L280]

[71]   Z. Ye, L. Lu, and M. Wong, "Zinc-oxide thin-film transistor with self-aligned source/drain regions doped with implanted boron for enhanced thermal stability", *IEEE Trans. Electron. Dev.,* vol. 59, no. 2, pp. 393-399, 2012.
[http://dx.doi.org/10.1109/TED.2011.2175398]

[72]   A. Valentini, F. Quaranta, M. Penza, and F.R. Rizzi, "The stability of zinc-oxide electrodes fabricated by dual ion-beam sputtering", *J. Appl. Phys.,* vol. 73, pp. 1143-1145, 1993.
[http://dx.doi.org/10.1063/1.354062]

[73]   W.S. Lau, and S.J. Fonash, "Highly transparent and conducting zinc-oxide films deposited by activated reactive evaporation", *J. Electron. Mater.,* vol. 16, pp. 141-149, 1987.
[http://dx.doi.org/10.1007/BF02655478]

[74]   R.L. Hoffman, B.J. Norris, and J.F. Wager, "ZnO-based transparent thin-film transistors", *Appl. Phys. Lett.,* vol. 82, pp. 733-735, 2003.
[http://dx.doi.org/10.1063/1.1542677]

[75]   S. Masuda, K. Kitamura, Y. Okumura, S. Miyatake, H. Tabata, and T. Kawai, "Transparent thin film transistors using ZnO as an active channel layer and their electrical properties", *J. Appl. Phys.,* vol. 93, pp. 1624-1630, 2003.
[http://dx.doi.org/10.1063/1.1534627]

[76]   P.F. Carcia, R.S. Mclean, M.H. Reilly, and G. Nunes, "Transparent ZnO thin-film transistor fabricated by rf magnetron sputtering", *Appl. Phys. Lett.,* vol. 82, pp. 1117-1119, 2003.
[http://dx.doi.org/10.1063/1.1553997]

[77]   J. Nishii, F.M. Hossain, and S. Takagi, "High mobility thin film transistors with transparent ZnO channels", *Jpn. J. Appl. Phys.,* vol. 42, no. 4, pp. L347-L349, 2003.
[http://dx.doi.org/10.1143/JJAP.42.L347]

[78]   B.J. Norris, J. Anderson, J.F. Wager, and D.A. Keszler, "Spin-coated zinc oxide transparent transistors", *J. Phys. D Appl. Phys.,* vol. 36, pp. L105-L107, 2003.
[http://dx.doi.org/10.1088/0022-3727/36/20/L02]

[79]   E.M. Fortunato, P.M. Barquinha, and A.C. Pimentel, "Wide-bandgap high-mobility ZnO thin-film transistors produced at room temperature", *Appl. Phys. Lett.,* vol. 85, pp. 2541-2543, 2004.
[http://dx.doi.org/10.1063/1.1790587]

[80]   S. Narushima, M. Orita, M. Hirano, and H. Hosono, "Electronic structure and transport properties in the transparent amorphous oxide semiconductor 2 CdO·GeO2", *Phys. Rev. B,* vol. 66, p. 035203, 2002.
[http://dx.doi.org/10.1103/PhysRevB.66.035203]

[81]   H. Hosono, M. Yasukawa, and H. Kawazoe, "Novel oxide amorphous semiconductors: Transparent conducting amorphous oxides", *J. Non-Cryst. Solids,* vol. 203, no. 8, pp. 334-344, 1996.
[http://dx.doi.org/10.1016/0022-3093(96)00367-5]

[82]   H. Hosono, "Ionic amorphous oxide semiconductors: Material design, carrier transport, and device application", *J. Non-Crystalline Solids,* vol. 352, no. 9-20, pp. 851-858, 2006.
[http://dx.doi.org/10.1016/j.jnoncrysol.2006.01.073]

[83]   Y. Shigesato, and D.C. Paine, "Study of the effect of Sn doping on the electronic transport properties of thin film indium oxide", *Appl. Phys. Lett.,* vol. 62, pp. 1268-1270, 1993.
[http://dx.doi.org/10.1063/1.108703]

[84]   P. Barquinha, A. Pimentel, A. Marques, L. Pereira, R. Martins, and E. Fortunato, "Effect of UV and visible light radiation on the electrical performances of transparent TFTs based on amorphous indium zinc oxide", *J. Non-Cryst. Solids,* vol. 352, pp. 1756-1760, 2006.
[http://dx.doi.org/10.1016/j.jnoncrysol.2006.01.068]

[85]   B. Yaglioglu, H.Y. Yeom, R. Beresford, and D.C. Paine, "High-mobility amorphous In2O3-10 wt % ZnO thin film transistors", *Appl. Phys. Lett.,* vol. 89, p. 062103, 2006.
[http://dx.doi.org/10.1063/1.2335372]

[86]   N.L. Dehuff, E.S. Kettenring, and D. Hong, "Transparent thin-film transistors with zinc indium oxide channel layer", *J. Appl. Phys.,* vol. 97, p. 064505, 2005.
[http://dx.doi.org/10.1063/1.1862767]

[87]   H.Q. Chiang, J.F. Wager, R.L. Hoffman, J. Jeong, and D.A. Keszler, "High mobility transparent thin-film transistors with amorphous zinc tin oxide channel layer", *Appl. Phys. Lett.,* vol. 86, p. 013503, 2005.
[http://dx.doi.org/10.1063/1.1843286]

[88]   W.B. Jackson, R.L. Hoffman, and G.S. Herman, "High-performance flexible zinc tin oxide field-effect transistors", *Appl. Phys. Lett.,* vol. 87, p. 193503, 2005.
[http://dx.doi.org/10.1063/1.2120895]

[89]   R.E. Presley, D. Hong, H.Q. Chiang, C.M. Hung, R.L. Hoffman, and J.F. Wager, "Transparent ring oscillator based on indium gallium oxide thin-film transistors", *Solid-State Electron.,* vol. 50, no. 3, pp. 500-503, 2006.
[http://dx.doi.org/10.1016/j.sse.2006.02.004]

[90]   K. Nomura, H. Ohta, K. Ueda, T. Kamiya, M. Hirano, and H. Hosono, "Thin-film transistor fabricated in single-crystalline transparent oxide semiconductor", *Science,* vol. 300, no. 5623, pp. 1269-1272, 2003.
[http://dx.doi.org/10.1126/science.1083212] [PMID: 12764192]

[91]   K. Nomura, H. Ohta, A. Takagi, T. Kamiya, M. Hirano, and H. Hosono, "Room-temperature fabrication of transparent flexible thin-film transistors using amorphous oxide semiconductors", *Nature,* vol. 432, no. 7016, pp. 488-492, 2004.
[http://dx.doi.org/10.1038/nature03090] [PMID: 15565150]

[92]   T. Iwasaki, N. Itagaki, and T. Den, "Combinatorial approach to thin-film transistors using multicomponent semiconductor channels: An application to amorphous oxide semiconductors in In-

Ga-Zn-O system", *Appl. Phys. Lett.,* vol. 90, p. 242114, 2007.
[http://dx.doi.org/10.1063/1.2749177]

[93] J.K. Jeong, J.H. Jeong, H.W. Yang, J.S. Park, and Y.G. Mo, "H. D. Kim HD, "High performance thin film transistors with cosputtered amorphous indium gallium zinc oxide channel", *Appl. Phys. Lett.,* vol. 91, p. 113505, 2007.
[http://dx.doi.org/10.1063/1.2783961]

[94] H.N. Lee, J. Kyung, and M.C. Sung, "Oxide TFT with multilayer gate insulator for backplane of AMOLED device", *J. Soc. Inf. Disp.,* vol. 16, no. 2, pp. 265-272, 2008.
[http://dx.doi.org/10.1889/1.2841860]

[95] J.K. Jeong, J.H. Jeong, and H.W. Yang, "12.1-in. WXGA AMOLED display driven by InGaZnO thin-film transistors", *J. Soc. Inf. Disp.,* vol. 17, no. 2, pp. 95-100, 2009.
[http://dx.doi.org/10.1889/JSID17.2.95]

[96] M. Kim, J.H. Jeong, and H.J. Lee, "High mobility bottom gate InGaZnO thin film transistors with SiOx etch stopper", *Appl. Phys. Lett.,* vol. 90, p. 212114, 2007.
[http://dx.doi.org/10.1063/1.2742790]

[97] C.H. Wu, H.H. Hsieh, C.W. Chien, and C.C. Wu, "Self-aligned top-gate coplanar In-Ga-Zn-O Thin-Film Transistors", *J Display Tech,* vol. 5, no. 12, pp. 515-519, 2009.
[http://dx.doi.org/10.1109/JDT.2009.2026189]

[98] D.H. Kang, I. Kang, S.H. Ryu, and J. Jang, "Self-aligned coplanar a-IGZO TFTs and application to high-speed circuits", *IEEE Electron Device Lett.,* vol. 32, no. 10, pp. 1385-1387, 2011.
[http://dx.doi.org/10.1109/LED.2011.2161568]

[99] J. Park, I. Song, and S. Kim, "Self-aligned top-gate amorphous gallium indium zinc oxide thin film transistors", *Appl. Phys. Lett.,* vol. 93, p. 053501, 2008.
[http://dx.doi.org/10.1063/1.2966145]

[100] B.D. Ahn, H.S. Shin, H.J. Kim, J.S. Park, and J.K. Jeong, "Comparison of the effects of Ar and H2 plasmas on the performance of homojunctioned amorphous indium gallium zinc oxide thin film transistors", *Appl. Phys. Lett.,* vol. 93, p. 203506, 2008.
[http://dx.doi.org/10.1063/1.3028340]

[101] S. Kim, and J. Park, "Source/drain formation of self-aligned top-gate amorphous GaInZnO thin-film transistors by NH3 plasma treatment", *IEEE Electron Device Lett.,* vol. 30, no. 4, pp. 374-376, 2009.
[http://dx.doi.org/10.1109/LED.2009.2014181]

[102] N. Morosawa, Y. Ohshima, M. Morooka, T. Arai, and T. Sasaoka, "Self-aligned top-gate oxide thin-film transistor formed by aluminum reaction method", *Jpn. J. Appl. Phys.,* vol. 50, no. 9R, p. 096502, 2011.
[http://dx.doi.org/10.7567/JJAP.50.096502]

[103] R. Chen, W. Zhou, M. Zhang, M. Wong, and H.S. Kwok, "Self-aligned indium gallium-zinc oxide thin-film transistor with phosphorus-doped source/drain regions", *IEEE Electron Device Lett.,* vol. 33, no. 8, pp. 1150-1152, 2012.
[http://dx.doi.org/10.1109/LED.2012.2201444]

[104] R. Chen, W. Zhou, M. Zhang, M. Wong, and H.S. Kwok, "Self-aligned indium-gallium-zinc oxide

thin-film transistor with source/drain regions doped by implanted arsenic", *IEEE Electron Device Lett.,* vol. 34, no. 1, pp. 60-62, 2013.
[http://dx.doi.org/10.1109/LED.2012.2223192]

[105] E.C. Knox-Davies, S.R. Silva, and J.M. Shannon, "Properties of nanocrystalline GaN films deposited by reactive sputtering", *Diamond Related Materials,* vol. 12, no. 8, pp. 1417-1421, 2003.
[http://dx.doi.org/10.1016/S0925-9635(03)00171-7]

[106] T. Maruyama, and H. Miyake, "Gallium nitride thin films deposited by radio-frequency magnetron sputtering", *J. Vac. Sci. Technol. A,* vol. 24, no. 4, pp. 1096-1099, 2006.
[http://dx.doi.org/10.1116/1.2208988]

[107] S. Zembutsu, and M. Kobayashi, "The growth of c-axis-oriented GaN films by dc-biased reactive sputtering", *Thin Solid Films,* vol. 129, no. 3-4, pp. 289-297, 1985.
[http://dx.doi.org/10.1016/0040-6090(85)90056-2]

[108] K. Kubota, Y. Kobayashi, and K. Fujimoto, "Preparation and properties of III-V nitride thin-films", *J. Appl. Phys.,* vol. 66, pp. 2984-2988, 1989.
[http://dx.doi.org/10.1063/1.344181]

[109] R.F. Xiao, H.B. Liao, N. Cue, X.W. Sun, and H.S. Kwok, "Growth of c-axis oriented gallium nitride thin films on an amorphous substrate by the liquld-target pulsed laser deposition technique", *J. Appl. Phys.,* vol. 80, pp. 4226-4228, 1996.
[http://dx.doi.org/10.1063/1.363302]

[110] S. Kobayashi, S. Nonomura, K. Abe, K. Ushikoshi, and S. Nitta, "Preparation of field effect transistor using nano-crystalline GaN", *J. Non-Cryst. Solids,* vol. 227-230, no. Pt2, pp. 1245-1249, 1998.
[http://dx.doi.org/10.1016/S0022-3093(98)00305-6]

[111] S. Kobayashi, S. Nonomura, and K. Ushikoshi, "Optical and electrical properties of nano-crystalline GaN thin films and their application for thin-film transistor", *J. Cryst. Growth,* vol. 189-190, pp. 749-752, 1998.
[http://dx.doi.org/10.1016/S0022-0248(98)00281-4]

[112] R.S. Chen, W. Zhou, and H.S. Kwok, "Top-gate thin-film transistors based on GaN channel layer", *Appl. Phys. Lett.,* vol. 100, p. 022111, 2012.
[http://dx.doi.org/10.1063/1.3676447]

[113] K.F. Mak, C. Lee, J. Hone, J. Shan, and T.F. Heinz, "Atomically thin MoS_2 : a new direct-gap semiconductor", *Phys. Rev. Lett.,* vol. 105, no. 13, p. 136805, 2010.
[http://dx.doi.org/10.1103/PhysRevLett.105.136805] [PMID: 21230799]

[114] B. Radisavljevic, A. Radenovic, J. Brivio, V. Giacometti, and A. Kis, "Single-layer MoS_2 transistors", *Nat. Nanotechnol.,* vol. 6, no. 3, pp. 147-150, 2011.
[http://dx.doi.org/10.1038/nnano.2010.279] [PMID: 21278752]

[115] S. Kim, A. Konar, W.S. Hwang, J.H. Lee, J. Lee, J. Yang, C. Jung, H. Kim, J.B. Yoo, J.Y. Choi, Y.W. Jin, S.Y. Lee, D. Jena, W. Choi, and K. Kim, "High-mobility and low-power thin-film transistors based on multilayer MoS2 crystals", *Nat. Commun.,* vol. 3, p. 1011, 2012.
[http://dx.doi.org/10.1038/ncomms2018] [PMID: 22910357]

CHAPTER 6

Driving Schemes and Design Considerations for AMOLED

Tsz Kin Ho[*]

Department of Electronic and Computer Engineering, The Hong Kong University of Science and Technology, Hong Kong

Abstract: State of the art thin-film transistor technologies continue to fuel new areas of researches and applications in flat-panel display. However, this does not come without new issues related to device-circuit stability and uniformity over large areas, placing an even greater need for new backplane designs, driving algorithms and compensation techniques in pixel architectures. In this work, we explained design considerations for active matrix backplane. We outline a systematic design approach, including circuit theory, enabling user to design circuits without worrying about the details of device physics.

Keywords: AMOLED, Backplane design, OLED design consideration, OLED driving.

CIRCUIT FUNDAMENTALS

In this chapter, we dedicated to circuit design techniques for thin-film transistor (TFT). In TFT technology, the dynamics of operation of transistors and circuits composed of transistors can be analyzed using approximate equivalent RC circuits. In this section, the RC circuits and the related dynamics are first introduced. Then we will go over the approximation process of transistor circuit in a systematic design approach. Finally, we will cover some special TFT circuit design techniques.

Resistor-Capacitor Circuit

The passive resistor and capacitor are the most fundamental circuit elements. The

---
[*] **Corresponding Author Tsz Kin Ho:** Department of Electronic and Computer Engineering, The Hong Kong University of Science and Technology, Hong Kong; Tel.: +852-23588845; E-mail: eeleo@ust.hk

resistance of a resistor, measured in Ohms, is also referred to as impedance. In integrated circuit design, metal thin films can be regarded as resistors. In metal thin films, the free electrons are accelerated when an electric field is applied. However, the accelerating electrons collide with other electrons and positive ions in the metal. The net effect leads to a constant velocity drift in the metal, producing current. By Ohm's Law, the current passing a resistor is directly proportional to the potential difference across the resistor, which are related as

$$V=IR$$

The passive capacitor consists dielectric insulating layer enclosed by two conductive electrodes can be treated as an energy storage device. The dielectric layer is insulating with very high resistivity, and thus charges cannot flow easily even under the driven of an electric field. However, with the present of electric field, positively charged nucleus and negatively charged electron move from their equilibrium position forming dipole, which develops charges stored onto the capacitor

$$Q=CV$$

A RC circuit is formed when these two components are connected in series. The RC circuit, which can be analyzed easily by first order rules, is often exploited to study the behavior of transistor circuit to obtain useful design information.

Charging and Discharging RC Circuit

When a RC circuit is connected to a voltage source, all components connected in series and so they share the same current throughout the process. In charging a capacitor, the initial current is the largest; this is because the initial voltage across the capacitor is zero, while the voltage across the resistor is maximum. As the process goes on, charges store in the capacitor that would develop a potential difference across the capacitor, and so potential across the resistor would decrease and hence the charging current decays. Fig. (**6.1**) shows the rate of change of the stored charges and charging current.

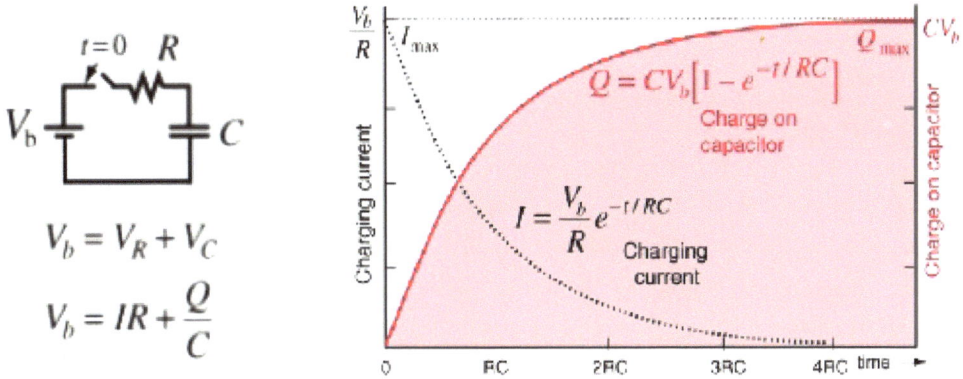

Fig. (6.1). Electrical characteristic in charging capacitor through a resistor [1].

Similarly, in discharging situation, now the capacitor itself acts as a voltage source. The voltage of both the capacitor and the resistor is the same throughout the process. From Fig. (**6.2**), the voltage, the discharging current and the capacitor's charge all follow the same type of decay curve during the process.

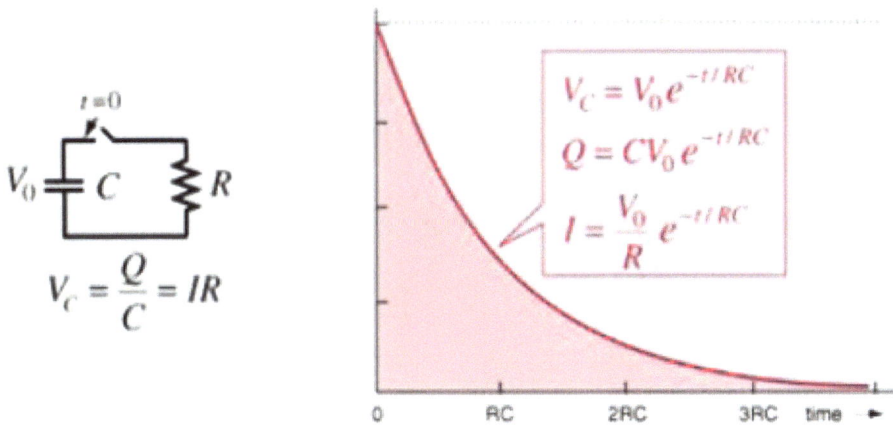

Fig. (6.2). Electrical characteristic in discharging capacitor through a resistor [1].

One important thing to note is the RC time constant, given by the product of the circuit resistance and the circuit capacitance. The rate of change of the above quantities can be described in terms of time constant. Ideally the capacitor takes infinite time to fully charge up, for our consideration, charging 90% of the applied voltage would equivalently take 2.3RC time. More reference, one RC, three RC and five RC time would correspond to a quantity reaches 63%, 95% and 99% out of total.

Capacitive Parasitics

Capacitive parasitic is an undesirable effect that is often called cross talk. [2] It is an unwanted coupling from a neighboring signal. The resulting perturbation serves as a noise source and it can cause intermittent errors, because the injected noise is determined by the transient value of the other signals routed in the neighborhood.

The potential impact of capacitive cross talk is influenced by the impedance of the line under investigation. The disturbance resulting from the coupling persists if the line is floating, and it becomes worsened when subsequent switching on adjacent wires. However, the signal can return to its original level if the wire is driven.

Let take an example from the configuration shown in Fig. (**6.3**). Assume that voltage at node X experiences a step changes equals ΔV_x, owing to the parasitic capacitance C_{XY}, this change will be coupled to node Y with amplitude given by the following equation,

$$\Delta V_Y = \frac{C_{XY}}{C_Y + C_{XY}} \Delta V_X$$

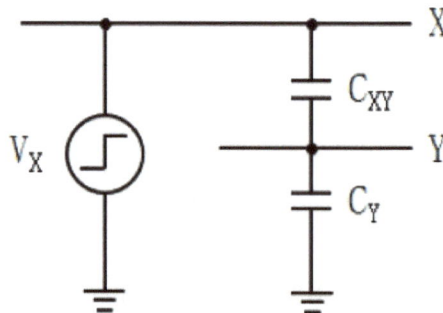

Fig. (6.3). Capacitive coupling to a floating line.

where C_Y is the self-capacitance inherited from node Y. This self-capacitance, sometimes referred as stray capacitance, varies depended on the size and geometry of the node. Circuits that are particularly susceptive to capacitive cross talk are network with pre-shared nodes, located in adjacency to full-swing wires. However, this effect can some-times be part of a key feature in TFT circuit design, such as bootstrapping circuit or special pixel design for specific purposes.

TFT CIRCUIT CONSIDERATIONS

Operational Region

Identifying transistor operational region is a fundamental, yet, crucial step for later TFT circuit analysis. This step is particular useful when we come to model transistor characteristic, and it serves as a foundation to all our assumptions made throughout modeling and design estimation processes.

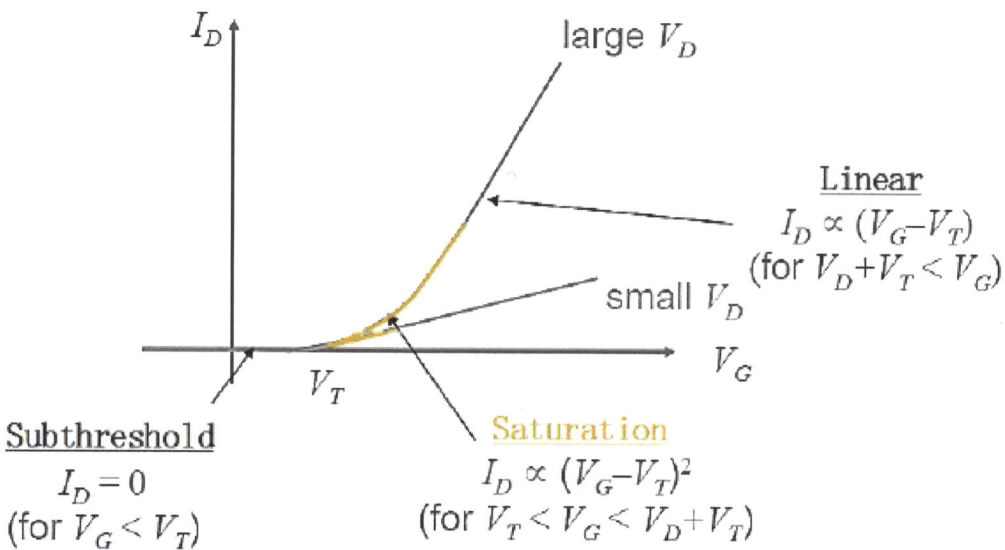

Fig. (6.4). Transistor operational regions for drain current (I_D) *versus* gate control (V_{GS}).

TFT in display circuit often works in digital mode, like an on/off switch, these transistors are usually called switching transistors. Switching transistors serve two purposes. Firstly, it isolates the source and drain nodes from interfering each other in off state. Secondly, it passes signal from one end to the other in swift manner.

That is to say, switch transistors have to be capable of flowing large current between their source and drain, these current would then charge or discharge capacitive node in their sub-sequence stage. Switching TFT are usually controlled by gate voltage that is provided from external driver chip, which able to provide large-voltage gate signal that is enough to bias the device to operate in linear region. This is further because linear current is larger than saturation current given a fixed V_{DS}. Fig. (**6.4**) and Fig. (**6.5**) shows schematic plot of transistor operational regions for drain current (I_D) and (log I_D) *versus* gate control (V_{GS}) respectively.

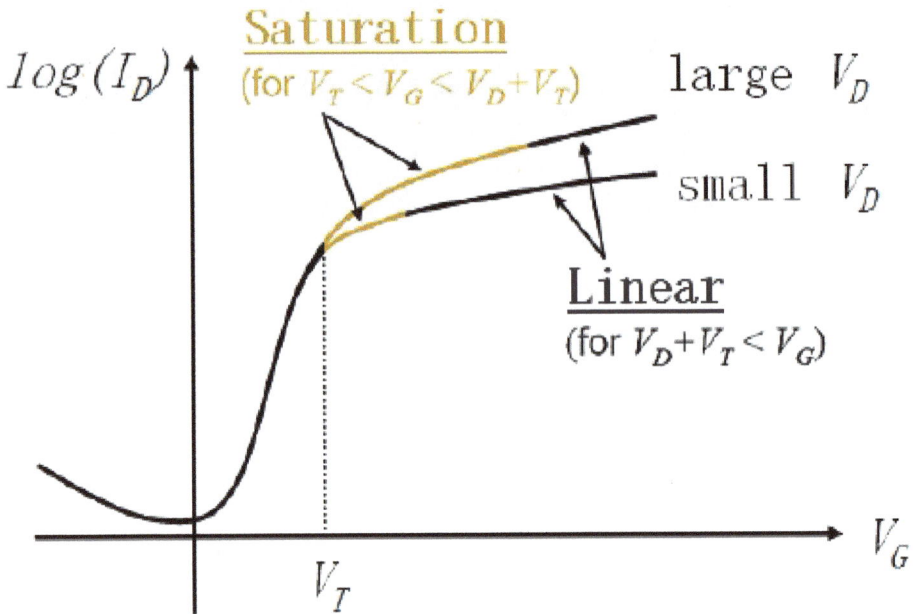

Fig. (6.5). Transistor operational regions for drain current log (I_D) *versus* gate control (V_{GS}).

Some devices in display circuit play another role, rather than as a switch. These tran-sistors often work in analog approach, for example, driving transistor that posits closest to the display element. Generally, there are at least one driving TFT in any display pixels. Such driving element usually is the largest device within the pixel due to through-put requirement for a relatively large loading associated from pixel electrode. Driving transistors work in saturation region, this is because saturation current provides good stability under fluctuation of V_{DS}. Let us take OLED as an example, OLED device is current sensitive, the drive transistor of

OLED pixel has to source stable and repeatable current according to data. This current is provided by the drive TFT in saturation, so that the current is purely depended on the gate voltage, which is the data, without other disturbance.

Transistor as a Switch

An electrical switch is one of the most fundamental elements of circuit. A TFT can be regarded as an electrical switch as shown in Fig. (**6.6**). The gate electrode controls the ON/OFF of the switch. When the switch is closed, the source and drain are connected because the channel becomes conductive. When the switch is open, the source is isolated from the drain because the channel is insulating.

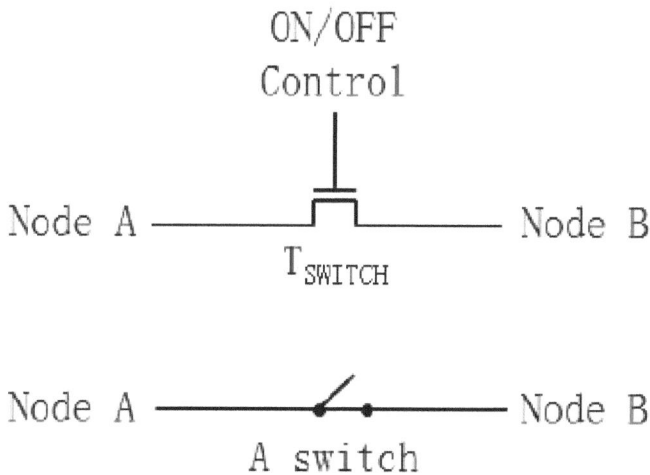

Fig. (**6.6**). Concept of a transistor as an electrical switch.

In ideal case, a perfect switch has no resistance when it is closed. And the resistance is infinite when the switch is opened. However in real case, there is always an on and off resistance associated to a closed and opened TFT switch respectively.

Transistor as a Current Source or Current Drain

Current source or Current drain is another common usage of transistor. A transistor can be a voltage controlled current source or drain as shown in Fig. (**6.7**). The gate terminal, instead of just being an ON/OFF control, modulates the amount of output current through V_{GS}. When a transistor is used as a constant current source or drain, it is essential to bias the transistor in saturation region. There are two reasons for that. Firstly, saturation current is less insensitive to the changes in V_{DS}. Secondly, the output impedance of transistor in saturation region is higher, so the output current from the source is less susceptible to changes despite changes in resistance value from the load. That is why an ideal current source has infinite resistance.

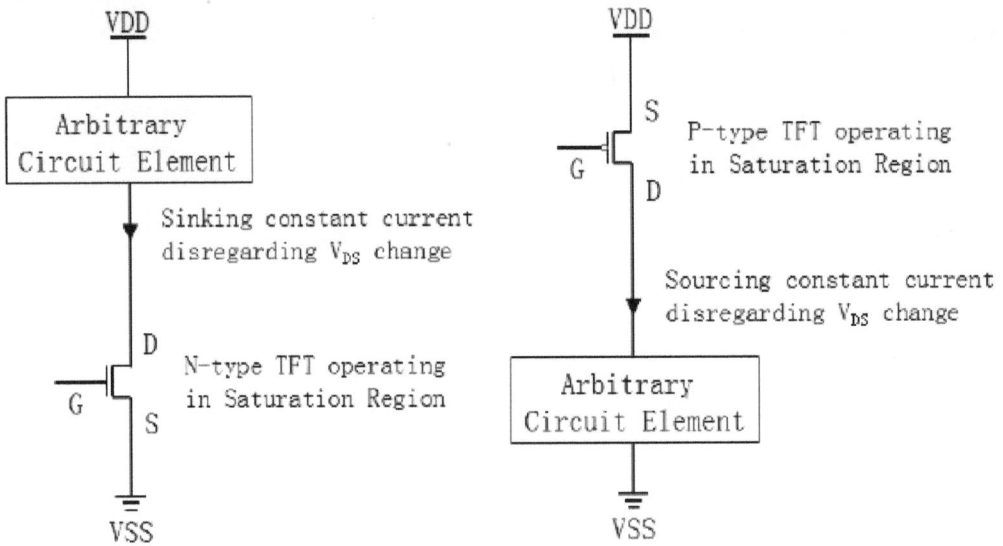

Fig. (6.7). An n-type TFT acts as current drain and p-type TFT serves as a current source.

In saturation region, P-type TFTs exhibit high output impedance and can be treated as good current source, while N-type TFTs behave as good current drain with high output impedance. Also, such configurations allow effective gate control (V_{GS}) since both of them are referencing to a DC power source.

On Resistance

In analog transistor circuit design, the small signal on resistance is often defined as inverse of the tangential slope under the I_D *versus* V_{DS} plot, mathematically expressed as

$$r_{ON} = \left(\frac{\partial I_{DS}}{\partial V_{DS}}\right)^{-1}$$

which is almost agreed in any semiconductor reference books. When it operates in linear region, the corresponding on resistance is the linear resistance or so called the output resistance. The on resistance in linear region can be given by the following expression,

$$r_{ON(Linear)} = \frac{1}{\mu C_{ox}\frac{W}{L}(V_{GS} - V_T - V_{DS})}$$

And when the transistor biased in saturation region, the corresponding small signal on resistance can be given by the following expression, it is also referred as the channel length modulation resistance,

$$r_{ON(CLM)} = \frac{1}{\lambda}$$

Fig. (**6.8**) depicted the graphical expression of corresponding on resistance of a transistor operate under linear region and saturation region. The red lines indicate the slope of the corresponding biasing point, and the resistance is given by the inverse of the slope.

On the other hand, TFT in display system including pixel and peripheral circuitry mostly operate in large signal manner, no matter which region it is biased in, the total output current does matter. Therefore, the analysis of TFTs' on resistance becomes directly equal to the fraction of V_{DS} over I_{DS}.

Fig. (**6.9**) sketched the on resistance for large signal analysis where the resistance is given by the inverse of the slope of red line. When the maximum operating V_{DS} (MAX) or sometimes V_{DD} is determined, we can estimate the value of R_{ON}.

Given a piece of transistor, the R_{ON} can be varied by V_{GS}. Fig. (**6.10**) demonstrated the difference in R_{ON} in compared to V_{GS}. When V_{DS} is constant, as V_{GS} increases, more inversion charges are being attracted to form the channel and effectively reduce the channel resistance, thus, R_{ON} drops.

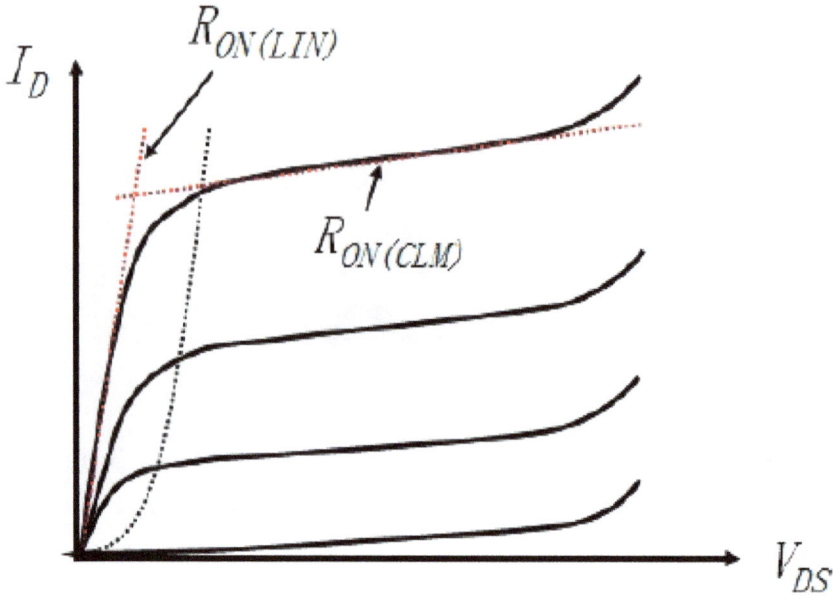

Fig. (6.8). On resistance at linear and saturation region for small signal analysis. The resistance value is the inverse of the slope.

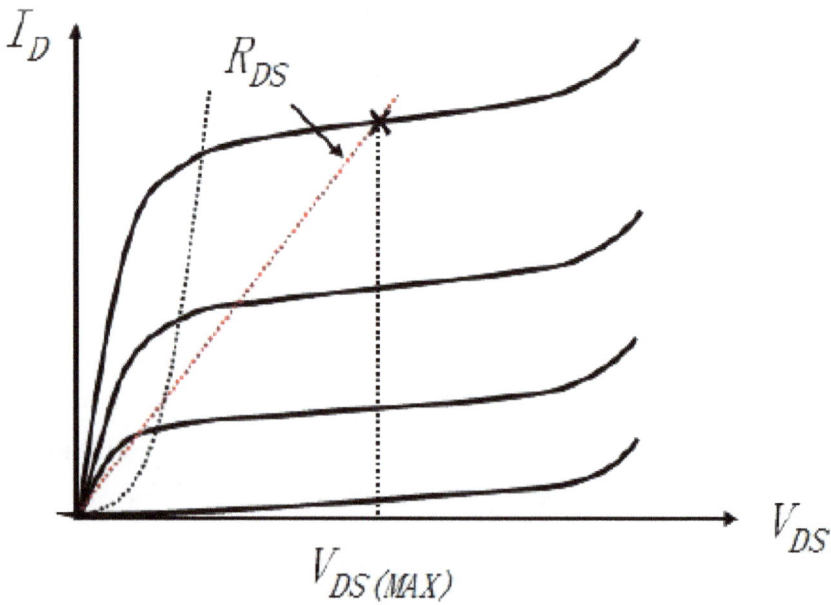

Fig. (6.9). On resistance for large signal analysis. The resistance value is the inverse of the slope.

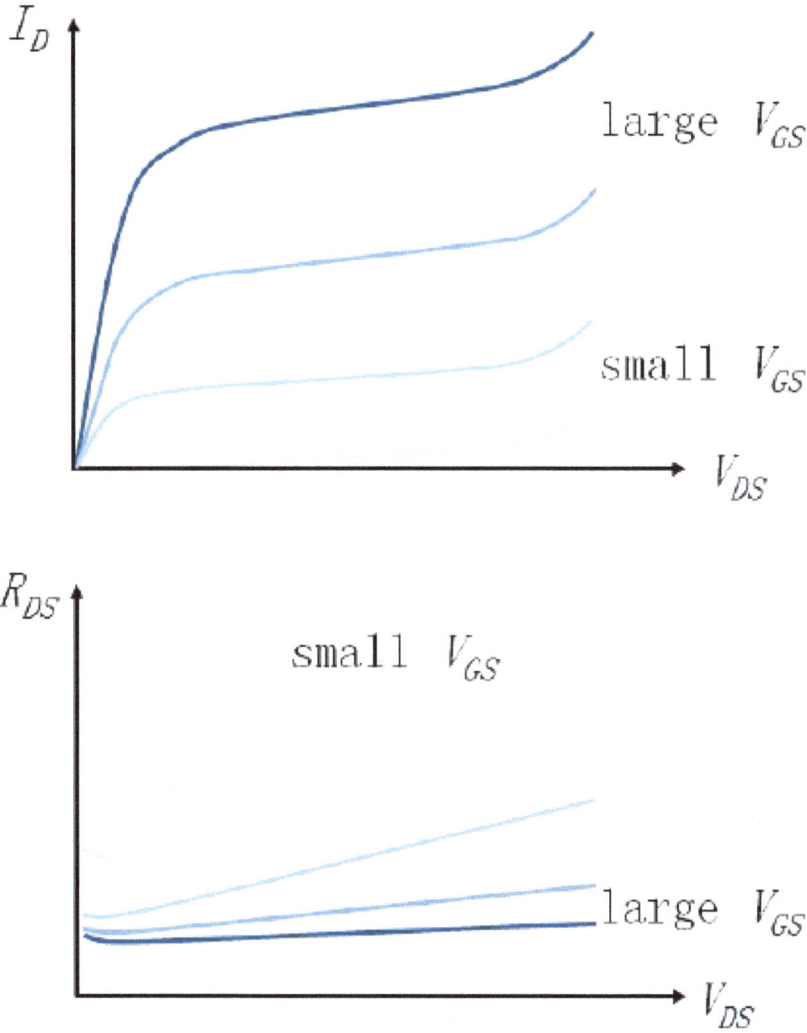

Fig. (6.10). Comparison of large-signal on resistance under different V_{GS} biasing.

Approximation of TFT with an equivalent resistance

In designing TFT circuit, it is a common practice to evaluate circuit performance by modeling transistor device as an equivalent resistance. The resultant circuit becomes RC network which simplifies the system and helps electronic engineers to achieve quick analysis. Although some degree of accuracy would be compromised, this approximation gives a reasonably accurate result; handful calculation steps would save a lot of time and computational effort in analysis.

When the operational voltage of a device is determined, the transistor characteristic can be approximated with an equivalent resistor ($R_{Equivalent}$). According to Fig. (**6.11**), the $R_{Equivalent}$ is located by the driven V_{GS} and the maximum operational V_{DS} (MAX), or it sometimes equals to V_{DD}. For different operation points, such as on state and off state, the equivalent resistance can be given by

$$R_{on} = \frac{V_{DS(MAX)}}{I_{DS(ON)}}$$

and

$$R_{off} = \frac{V_{DS(MAX)}}{I_{Leakage}}$$

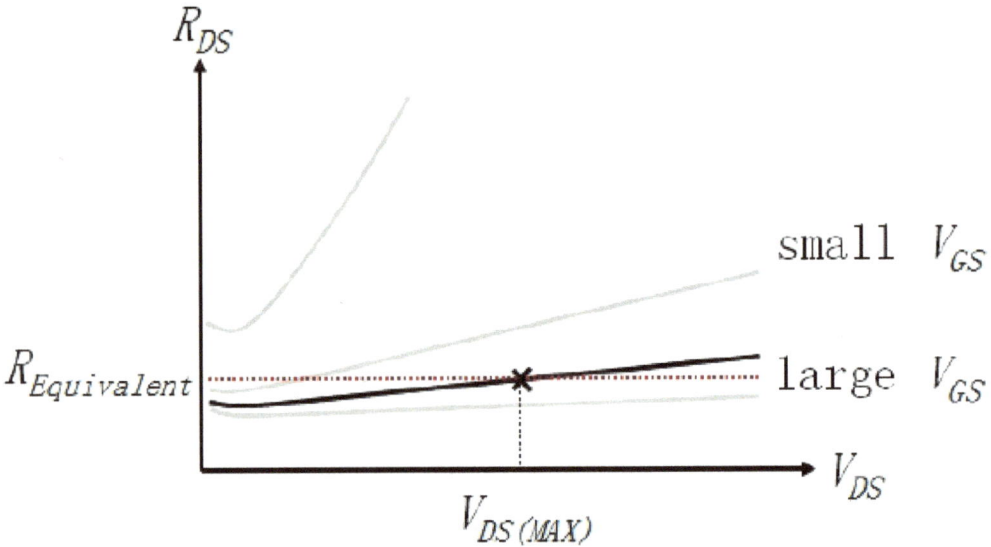

Fig. (**6.11**). Approximate transistor output resistance.

where $I_{DS(ON)}$ and $I_{Leakage}$ is the drain current biased at $V_{GS}(ON)$ and $V_{GS}(OFF)$. We can treat this approximation as for the worst case scenario. This is because $R_{Equivalent}$ is always larger than the actual transistor's output resistance as V_{DS} decreases. And the accuracy of such approximation considerably increases as V_{GS}

increases, since the variation of R_{DS} drops significantly when V_{GS} is high. Therefore, this approximation is particularly useful for TFT switches, which is operated in deep linear region with $|V_{GS}|$ normally, go very high. For example, in mobile applications, the transistor gate voltage swing is usually between 15 V to 30 V; while for stationary display, this voltage swing can go up to 70 V depending on the transistor technology and transfer characteristics.

We can verify this approximation by the following simulation. Firstly, we use ELA TFT technology, which has transfer characteristics as shown in Fig. (**6.12**), as an example. Secondly, the transistor operation point for $|V_{GS}|$ and $|V_{DS}|$ are 11.3 V and 5 V respectively. The corresponding $|I_{DS}|$ is about 230 uA. Given this information, the approximated $R_{Equivalent}$ is about 21.7 KΩ.

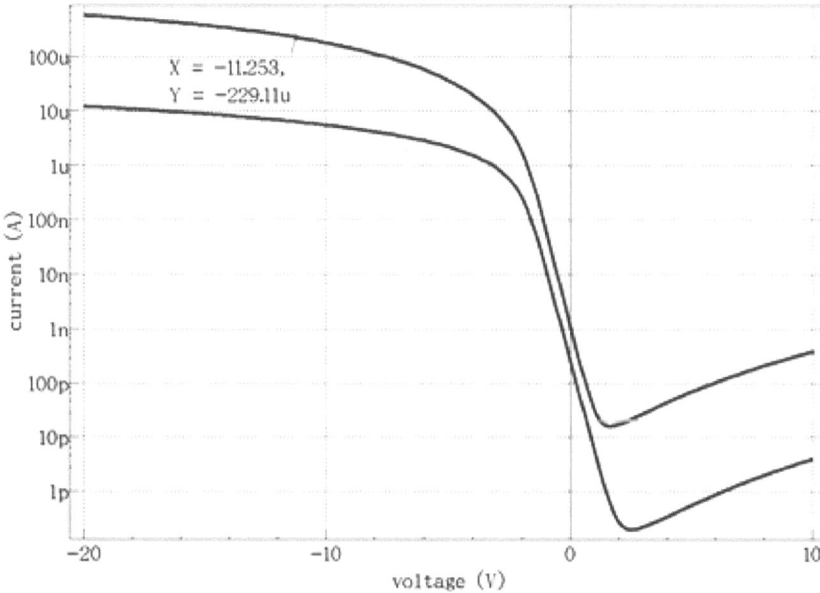

Fig. (6.12). Transfer characteristics of ELA TFT process showing $|I_{DS}|$ at $|V_{DS}|$ equals 0.1V and 5V.

Consider we charge up a capacitor of 1 nF with the TFT and the $R_{Equivalent}$ independently. The biasing point was designed such that the charging process completes in 100 us. The transient capacitor voltage (V_C) of both cases is simulated and compared. The circuit schematics and simulation result are shown in Figs. (**6.13 and 6.14**).

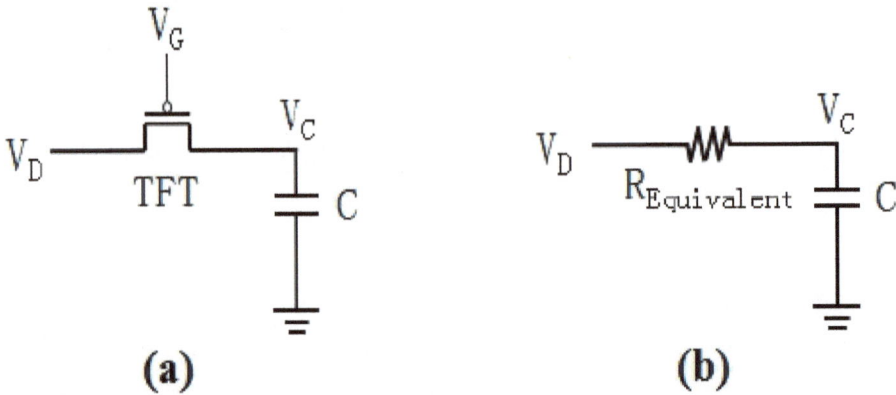

Fig. (6.13). (a) Circuit of TFT switch-capacitor, (b) the TFT switch can be treated as an equivalent resistance.

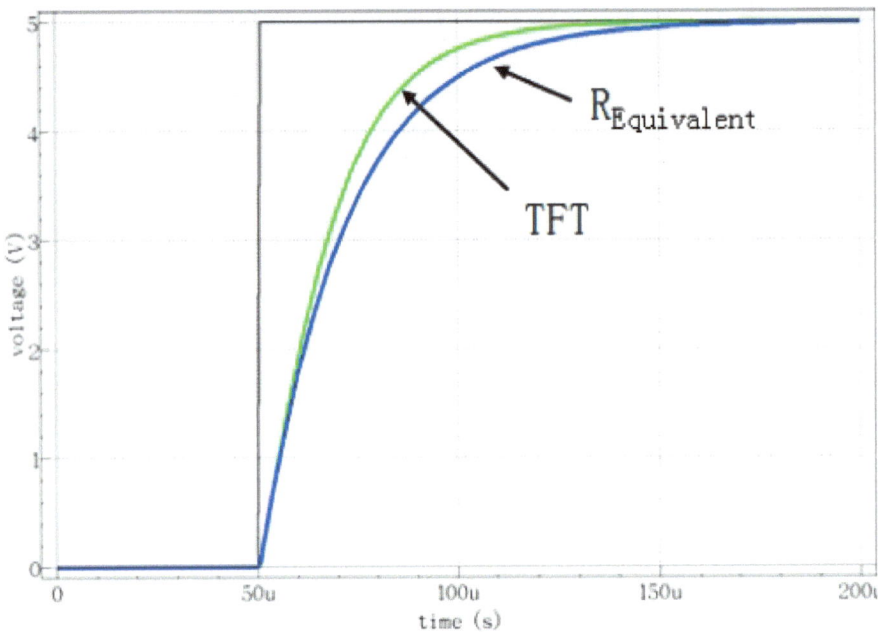

Fig. (6.14). Transient simulation result of charging up a switch capacitor circuit with TFT and equivalent resistance.

The $R_{Equivalent}$ circuit is the worst case approximation, therefore, the capacitor voltage shown a faster dynamic in the TFT circuit. Notice that, both cases meet the timing requirement of charging up 99% of the capacitor in 100 us, and both cases show consistent results with less than 1% difference.

DESIGN CONSIDERATIONS FOR ACTIVE-MATRIX BACKPLANE

The consideration of a designer has to be thorough and organized. Even the slightest mistake made in a design may affect product performance, functionality or, in worst case, may destruct design operation and end up bringing failure to the project. Similar to other display circuit, transistors in OLED pixel can be categorized into two groups, switching transistors and driving transistors. Recall the 2T1C basic circuit, transistor M1 mainly function as a digital switch and M2 act as a current driver or current source to the OLED. Therefore, the approaches in designing these two transistors are different. Some design requirements and considerations mainly affect the switching transistors, and vice versa. By taking 2T1C as example, we will discuss all these design considerations one by one.

Brightness

One of the performance measures for an OLED display is brightness. It is characterized by the display output luminance which equals to the amount of luminous flux spread over a given area with a unit of candela per square meter (cd/m2), and sometimes also called nits. Larger this parameter would equivalent to a brighter source. And yet, the luminance requirement is depended on the applications and the usage conditions, the luminance requirement of different applications have been shown in Table (**6.1**). For examples, outdoor displays usually require higher brightness than indoor displays in order to improve sunlight readability; while the brightness requirement for indoor display is lower, such as PC monitor and television. In fact, sometimes we want to avoid high using bright screen, this is because prolonged usage of bright screen would lead to unpleasant feeling like eyestrain and fatigue. For portable devices, high brightness display increases the energy consumption rate, which reduces the battery life.

Table 6.1. The luminance requirement of different applications.

Applications	Required luminance (Cd/m²)
Mobile Phone	500
Monitor	80-450
Television	450-1000

Depending on the applications or usage condition of the display, we can define the required luminance of an OLED display as design target and estimate the required pixel current using the OLED characteristic. Fig. (**6.15**) shows the relation between luminance, current density and the applied voltage across an OLED test device.

Assume we have a design target that our OLED panel requires to meet a maximum luminance of 500 cd/m^2. From Fig. (**6.15**) we can find out the corresponding voltage is about 6.5 V and the respected current density is 3 mA/cm^2.

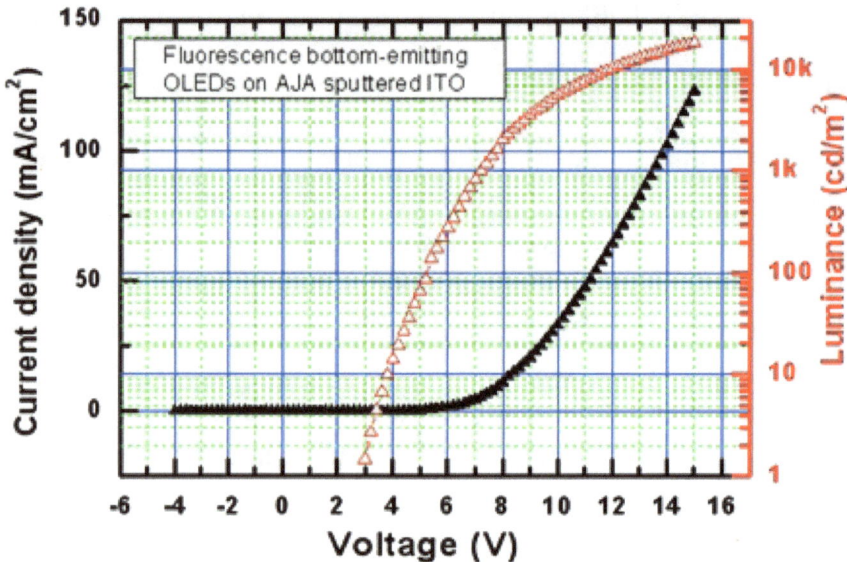

Fig. (6.15). OLED Current Density and Luminance Characteristics.

For each pixel, let say, the OLED emission region is 0.00007 cm^2, the corresponding maximum OLED current required will be 0.2 uA. This is the current required to be generated from an OLED pixel in order to give the maximum display luminance. With these parameters, we can plan out the OLED pixel design to ensure it supply the needed voltage and current to the OLED devices.

Display Timing

Timing is another major design issue in AMOLED display. It related to the display refresh frequency, also called the frame rate, and the display resolution. Conventional displays update at high speed in order to reduce image sticking, the minimum frequency required to avoid flickering is 60Hz. Meaning that, a display is flickering-free to users only when it generate no less than 60 sub-frames every seconds. For 3D applications, this requirement goes even higher, like 360Hz, to ensure display smoothness to viewers.

Let us first limit our discussion to a conventional display with resolution is 1920 x 1080 for Full HD 1080P operating at 60 Hz. Meaning that there are 1080 horizontal lines and 1920 vertical columns, and the scan bus and data bus lie horizontally and vertically in between each pixels respectively, forming a rectangular matrix. These signal bus are shared among the rows and columns in order to drive the panel progressively. We can easily work out the time limit to update a row of pixels (T_{ROW}),

$$T_{ROW} = \frac{1}{\text{frame rate} \times \text{horizontal lines}} \approx 15.4us$$

The above equation explains that there will be 60 sub frames generated per second, and each frame consists of 1080 lines, therefore, the display has to complete data update for a row of pixel within 15.4 us. And the data update action take place in progressive scanning manner, which means the update process scans through the display row-by-row sequentially. After data update, a pixel has to hold and output according to the data for the rest of the sub frame before the next update. Therefore, the data update time, also called the scan period, is 15.4 us. While the data holding time, or so called the frame period, is 16.7 ms.

Pixel Storage Capacitance

Generally speaking, all active matrix display pixel need to have a memory unit called the pixel storage capacitor, no matter it is for LCD display or OLED display. This memory unit is periodically charged to a voltage level corresponding to the desired gray level of the pixel. In order to maintain the data voltage within adequate level during the frame period, the capacitance value and leakage current are the most important.

The total pixel storage capacitance is a lumped capacitance with artificial and parasitic capacitance. Parasitic capacitance could be due to different components of the pixel such as the liquid crystal cell capacitance in LCD's pixel or the gate capacitance of driving transistor in OLED's pixel. Artificial capacitor can be constructed by overlaying two or more conductive layers with a dielectric layer sandwiched in between, the structure of this capacitor is simple and easily controlled. For bottom emitting AMOLED, however, the presence of such capacitance will reduce the pixel aperture. This is because most conductive layers are opaque and the capacitance value depends on the overlapping area. Therefore, larger capacitance takes up more space and it will limit pixel opening and block the light emission of pixel.

The pixel storage capacitance has to be carefully designed. If the designed pixel capacitance was too small, the pixel storage cannot effectively hold the pixel data within the frame period; while if the pixel capacitance was too large; there could be insufficient time to charge this capacitor to the desired data level and suffering unnecessary loss in pixel aperture. Both of these cases affect pixel data integrity, impairing brightness and display contrast. Either cases point out the importance of finding the adequate value of the pixel storage capacitance.

Design Expression

In TFT circuit design, we often come across situations like designing size of transistor or capacitance for pixel circuit or other peripheral circuit, such that a circuit can meet the operational requirement within specific period of time. Generally, this situation can be divided into two types; they are RC charging process or RC discharging process. Let us first consider the RC charging process, the equations of charge on a capacitor at any given time can be expressed as

$$Q(t) = CV_o\left[1 - e^{-\frac{t}{RC}}\right]$$

where V_o is the target charging voltage, C and R is the capacitance and resistance in the RC network respectively. Assume we are given a charging ratio (k), which defines the ratio of V_o to be developed in the capacitor by the end of the charging operation, because a capacitor takes infinite time to reach the target charging voltage. And we are given a time (t) that specify the time limit for this charging operation. For design convenience, we can further elaborate the above equation as followed

$$Q(t) = Q_{max}\left[1 - e^{-\frac{t}{RC}}\right]$$

$$e^{-\frac{t}{RC}} = 1 - \frac{Q(t)}{Q_{max}}$$

$$-\frac{t}{RC} = ln(1 - k)$$

therefore,

$$R_{max} = -\frac{t}{C \cdot ln(1 - k)}$$

In the above equation, Rmax is equivalent to the maximum resistance allowed to fulfill the stated requirement in the RC circuit. This can then be devised into the minimum required current (Imin) in achieving such requirement, it can be given by

$$I_{min} = \frac{V_{DS(max)}}{R_{max}}$$

With the minimum required current (I_{min}), it is much easier to estimate the respected TFT size through table lookup or interpolation.

Similar approach can be applied to RC discharging process, this time, assume we have to hold a voltage V_o within a given discharging ratio (k) for a period of time (t). Starting with the equation of charge on a discharging capacitor at any given time, we have

$$Q(t) = CV_o \cdot e^{-\frac{t}{RC}}$$

where R can be given by the approximated off resistance of the transistor circuit like this

$$R = \frac{V_{DS(max)}}{I_{Leakage}}$$

Rearranging the equation of charge, we have

$$\frac{Q(t)}{Q_{max}} = e^{-\frac{t}{RC}}$$

$$ln(k) = -\frac{t}{RC}$$

$$C_{min} = -\frac{t}{R \cdot ln(k)}$$

In the above equation, C_{min} is equivalent to the minimum capacitance required to fulfill the requirement.

These expressions are particularly useful when doing the first order estimation. By analyzing the simple RC circuit, the time constant of the dynamics can be easily identified. This facilitates the designer to make quick calculation when they design the capacitance of the capacitor and aspect ratio of the TFT to achieve the desired time constants.

TFT CIRCUIT DESIGN TECHNIQUES

Bootstrap Circuit

The word bootstrapping is defined as to lift oneself by the bootstraps to achieve success by one's own unaided efforts. A similar phenomenon may occur in some electronic circuits, especially for TFT system-on-glass design. Due to yield rate consideration of large sized display, TFT manufacturing process would target for simplicity and uniformity. Unlike conventional CMOS circuit, in which N-type MOSFET and P-type MOSFET constructs pull-down network and pull-up network respectively, displays manufacturing process usually comes with only single typed transistors. Meaning that pixel and peripheral circuit composed of only N-type or P-type transistor for both networks. Lack of a complementary pull-up or pull-down network, a pseudo N-type or P-type design cannot achieve rail-to-rail signal swing at output stage due to threshold drop.

Bootstrapping is generally applied to the missing network of the circuit. Capacitive coupling is intentionally added, Using signal transient, the coupling effect would boost up voltage level of one node above or below normal operation range, such signal would instantaneously drive transistors of output stage to compensate the threshold lost due to the missing type.

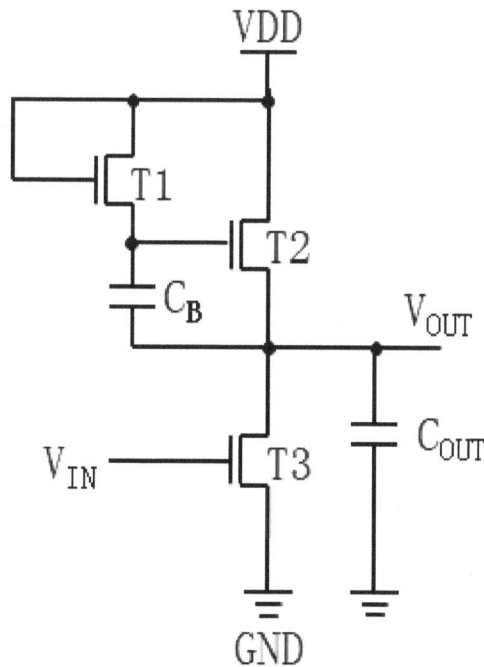

Fig. (6.16). Bootstrapped N-type inverter.

Such mechanism can be illustrated in the above bootstrapped N-type inverter with schematic as shown in Fig. (**6.16**). The circuit is constructed by merely N-type transistors, which are best working as pull-down network. When input is high, T3 will be turned on, output (V_{OUT}) will be discharged to GND. On the other hand, when input switched to low, T3 will be off. Transistor T2, due to the diode-connected transistor T1, will charge up the output up to V_{DD}-V_{TH}. The bootstrapping capacitor (C_B) will couple this voltage change to the other end that is connected back to the gate of T2. The bootstrapping effect pushes in the gate of T2 to a potential greater than V_{DD}+V_{TH}, keeping T2 operates in the linear region. The output is hence able to reach V_{DD}.

CIRCUIT COMPENSATION AND LAYOUT DESIGN

OLED pixel is the major basic unit of an OLED display. Although there are some applications of passive matrix OLED display, the mainstream lies into active matrix application thanks to the sophisticated development of TFT industry over recent decades. The major differences between OLED display and LCD are their light emitting mechanism. Unlike its counterpart, OLED is a light emitting device which operates in a stand-alone manner without the need for the backlight unit. OLED is a current driven devices, meaning that the light intensity of an OLED output depends on the amount of current flowing through this device. Due to its diode-like property, it is often to use a diode to indicate or to model the operation of an OLED in circuit simulation. In this section, we will first give an introduction of the development in AMOLED display. Then, we will discuss about some issues arising from TFT threshold voltage shifting in AMOLED display. After that we will present the method of threshold voltage compensated OLED pixel and simulation for different configurations.

CHALLENGE IN AMOLED DISPLAYS

Aging of OLED and TFT

Display lifetime is when the display luminance has dropped to half of its initial value. This is usually caused by the effect of OLED luminance degradation and TFT degradation. For a smaller display, the lifetime is smaller, say about 3000 hours. On the other hand, for larger displays, the required lifetime is about 50000 hours. To minimize the effect of aging, compensation scheme is applied to manage this effect, TFT is given a headroom between maximum overdrive voltage and the operating voltage V_{DD}. Also, the OLED lifetime increases as the aperture ratio increases. This is because the OLED degradation is a function of the current density [3 - 5], larger aperture effectively reduce the current density for a given brightness. However, the aperture ratio is limited by the size and number of transistors we have in a pixel, therefore, achieving the required lifetime in the least pixel complexity can be challenging.

Threshold Voltage Shift

Despite the fact that OLED would lead to a more compact and a better contrast display in comparison to LCD, there are another well-known issues associated with OLED displays, which is the threshold voltage shifting of thin-film transistors. This effect can be resulted by a number of factors including charge trapping by interface states and oxide traps, drift of mobile ions and self-heating [6 - 10]. Fig. (**6.17**) shows an example of threshold voltage shift of a transistor over time. Due to prolonged electrical stress, the threshold voltage of a transistor may deviate about 1 V in a few days. Without proper compensation measure, OLED displays would show clouding-like defects, which is also known as mura, as shown in Fig. (**6.18**). This becomes the challenge of having threshold voltage compensated active-matrix OLED pixel.

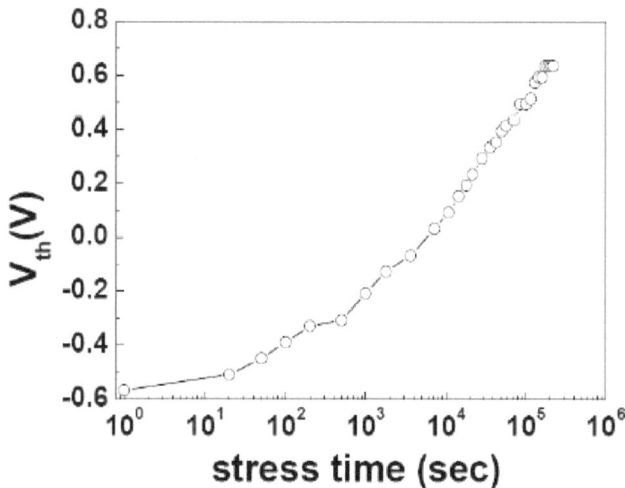

Fig. (6.17). Threshold voltage shift of a poly-silicon TFT subjected to hot carrier stress ($V_G=V_D/2$, $V_D=14V$) [9].

Fig. 6.18. Mura effect in OLED display without threshold voltage compensation.

2T1C Pixel Configuration

2T1C configuration is the most basic driving method in AMOLED. Such pixel design consists of two transistors and one capacitor. The schematic arrangement of 2T1C OLED pixel is shown in Fig. (**6.19**). The diode (D1) depicts a modeled OLED device in the pixel. And the red arrow labels the current path flowing through it during illumination. Since p-type transistors work best as current source, OLED pixels mostly employ p-type devices to build the circuit.

There are two signals and two powers connected to this pixel, DATA, SCAN, V_{DD} and V_{COM} respectively. The working principle of 2T1C pixel is simple. During the scan cycle of the pixel, SCAN will be asserted and the data signal will be ready in DATA at the same time.

The data signal will be coupled and stored in the capacitor (C1) for the whole period of frame duration. After the scan cycle, the SCAN signal switch off transistor M2, while the current flows through transistor M1 and OLED device (D1) is controlled by the voltage storage within C1. And this current can be equated as follow.

$$I_{OLED}=K(V_{DD}-V_{DATA}-V_{th})^2$$

where $K = \dfrac{\mu C_{ox} W}{2L}$ is a parameter depends on the physical characteristics of transistor M1. We can see the OLED current depends not only to the data voltage, but also the threshold voltage of M1. And this threshold voltage will be changed over time due to electrical stress. This arise the issue of display non-uniformity by cause of threshold voltage shift. To show the severity of issue due to threshold voltage shift, a simulation has been made. Assume we have a 2T1C pixel fabricated in ELA TFT technology with IV characteristic.

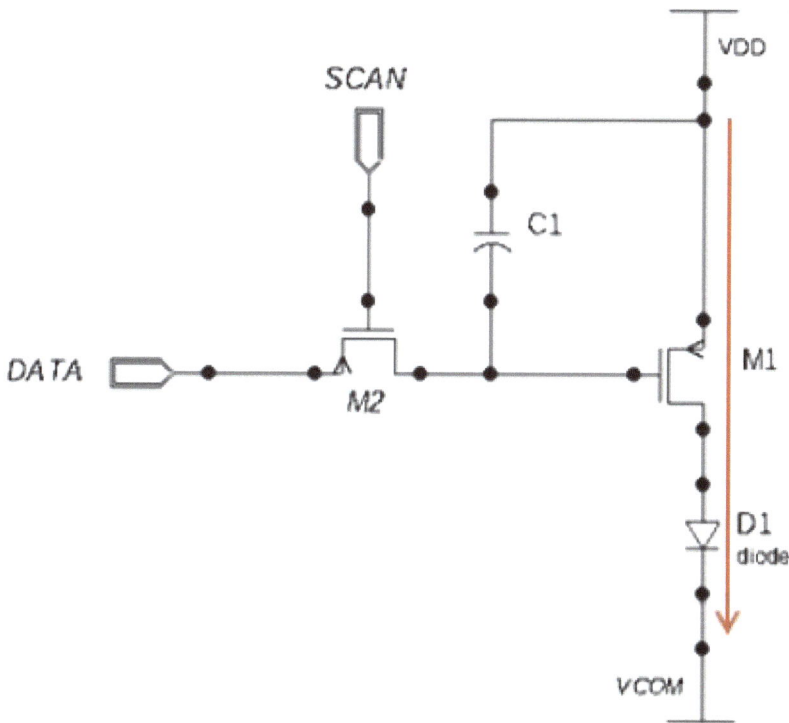

Fig. (6.19). Schematic of conventional 2T1C OLED pixel.

Fig. **(6.20)** shows the simulation result of current change against threshold voltage shift when the pixel current is 400 nA. We can see that when ΔV_{TH} is zero, the pixel current as expected is 400 nA, so the change in the pixel current is zero, showing the case where there is no threshold voltage shift initially. However, when the threshold voltage increase to about 0.6 V, the pixel current is reduced by half; or when the threshold voltage reduced by 2 V, the pixel current is almost vanished. Note that the ΔV_{TH} here indicates the value change of a P-type transistor.

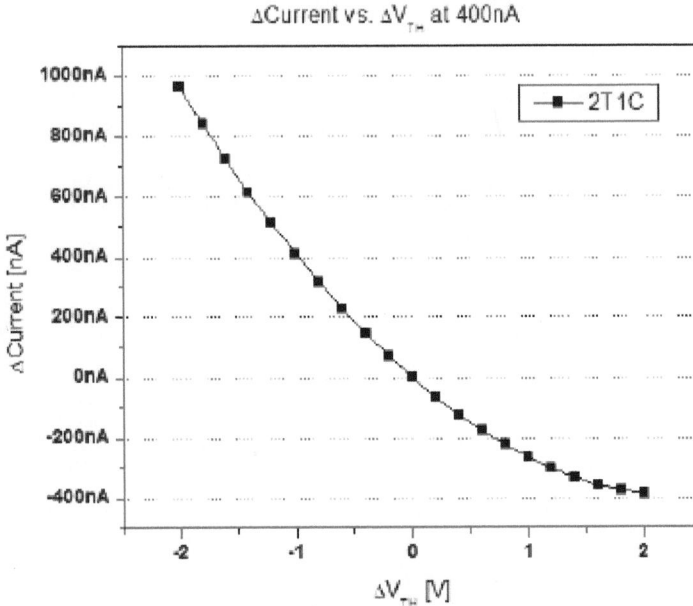

Fig. (6.20). Simulation result of current change against threshold voltage shift in 2T1C OLED pixel when the pixel current is 400 Na.

THRESHOLD VOLTAGE COMPENSATED AMOLED PIXEL

3T1C Pixel Configuration

The 3T1C pixel was designed to compensate for the shift of device parameters that are exist in amorphous TFT. However, it cannot compensate for parameter variation from pixel to pixel. Fig. (**6.21**) depicted the schematic design of such pixel.

The basic operation principle is similar to the conventional 2T1C design. During addressing time, transistor M3 is closed and the data voltage is stored in the capacitor C1. The current passing through transistor M1 is controlled by the data voltage. Current drifts can be compensated by transistor M2 which serves as an active resistor. If current passing through M1 changes, then the voltage across the M2 will change accordingly. For instance, if the current decreases, the voltage across the M2 decreases as well, and thus higher current can flow to the OLED because the voltage of the capacitor stays constant. The current provided to the OLED devices is given by

$$I_{OLED}=K(V_{DD}-V_{DATA}-V_{th(M2)}-V_{th(M1)})^2$$

As mentioned before, the merits of this configuration is simplicity, as there are only one extra transistor added to the conventional design and no additional control signal required, therefore, it favors panel fabricated in amorphous TFT or bottom-emitted OLED where transistors area considerably affect pixel aperture. However, this is an indirect method of threshold voltage shift compensation.

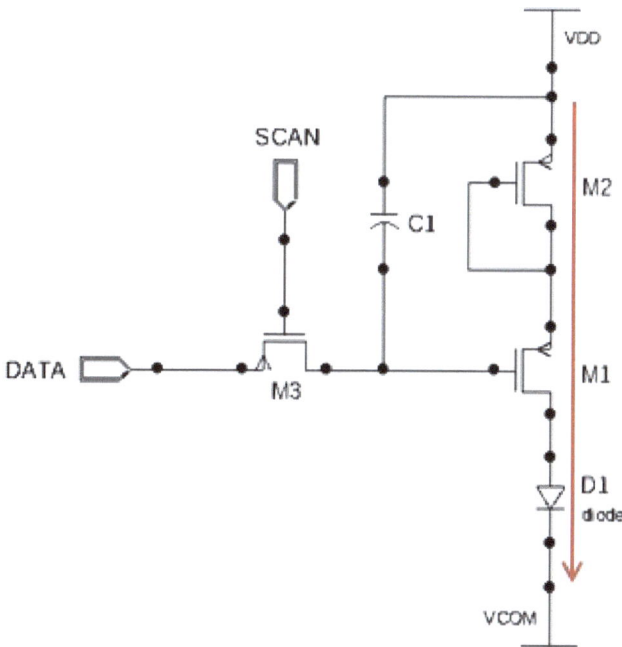

Fig. (6.21). Schematic of threshold voltage shift compensated 3T1C OLED pixel.

This method, to some extent, slows down the effect of threshold voltage shift of each pixel individually, rather than compensates the effect. The point is there is no guarantee between the threshold shift and the voltage drop across the active resistor. The current lost due to the shift could not be achieved back to the original level through this method, but decreased in a milder extent. Moreover, both transistor M1 and M2 suffer electrical stress during pixel operation that means both devices undergo threshold voltage drift, and that effectively reduce the compensation ability of the pixel. Fig. (**6.22**) shows the simulation result comparing 3T1C and 2T1C OLED pixel. In the result, the 3T1C pixel has lower variation of current than the 2T1C pixel. However, both designs are still far away to be called reliable.

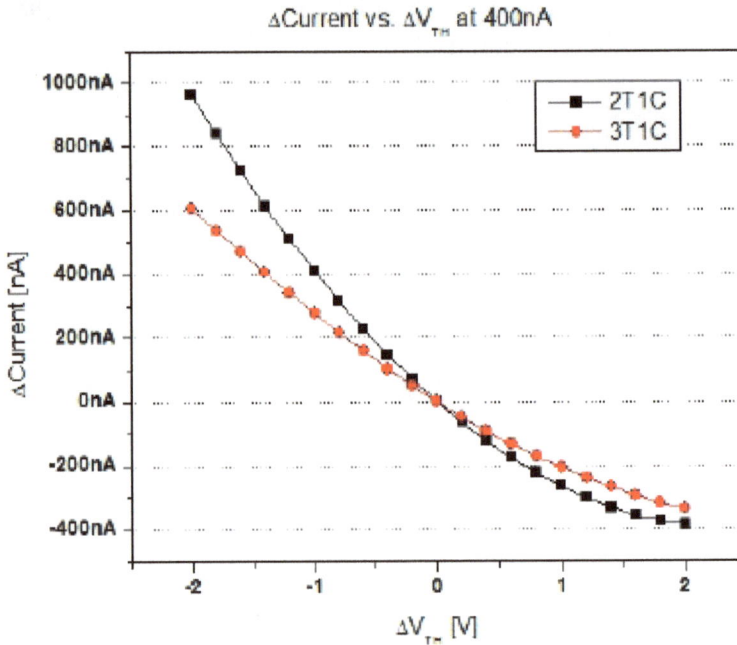

Fig. (6.22). Simulation result of current change against threshold voltage shift in 3T1C OLED pixel when the pixel current is 400 nA.

4T1C Pixel Configuration

Current driven pixel architectures are attractive for AMOLED display, because of their ability to tolerate threshold voltage shift, mismatches and non-uniformity caused by aging. A 4T1C pixel is one of the current driven OLED structures developed from LG Philips LCD and Korea University [11]. Fig. (**6.23**) shows the schematic structure of current driven 4T1C OLED pixel.

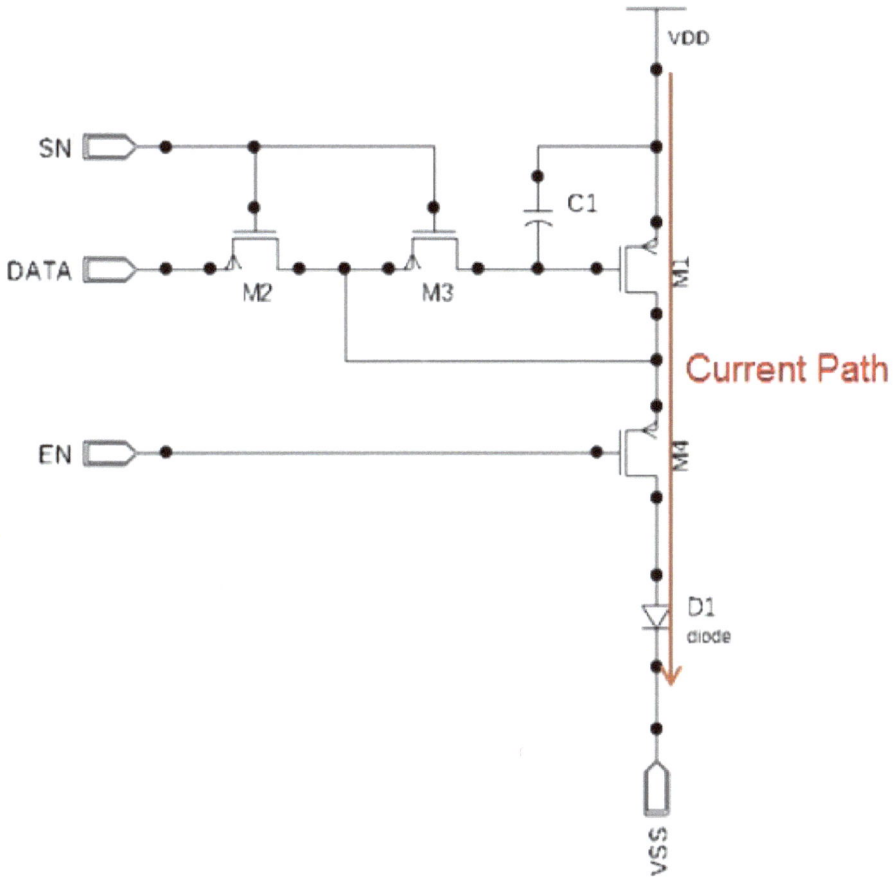

Fig. (6.23). Schematic of current driven 4T1C OLED pixel.

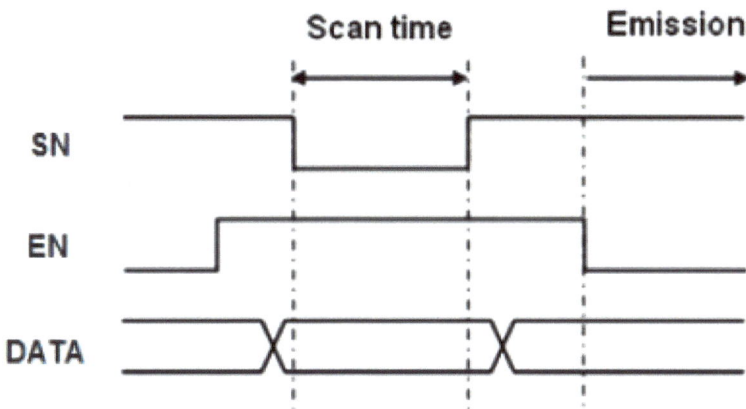

Fig. (6.24). Driving waveform of current driven 4T1C OLED pixel.

Fig. (**6.24**) shows the driving waveform for the 4T1C architectures. The operation of the 4T1C pixel will first started by the scan phase, and because it is current driven, the data signal is a current. The scan pulse (SN) will turn on transistors M2 and M3, allowing the data current to flow from V_{DD}, passing M1 and M2, to the data driver. Transistor M3 diode connected M1, forcing M1 to operate in saturation region. At the same time, the overdrive voltage (V_{GS}) developed from M1 is stored in capacitor C1.

In emission phase, scan signal (SN) switch off M2 and M3. While enable signal (EN) switch on M4 and let current flow to the OLED device. This current is reproduced by the overdrive voltage stored inside C1 with the same amplitude as the input data disregarding threshold voltage shift of any devices. Fig. (**6.25**) shows the relation of input and output that almost perfectly matched. However, the tradeoff for using current driven pixel scheme is the non-uniform data settling time. Unlike charging and discharging of RC circuit, the scan operation here is undergoing a constant current charging or discharging process, which means given a capacitance, the required completion time depends on the value of data current. Fig. (**6.25**) shows the exponential relation between the required transient and the input current.

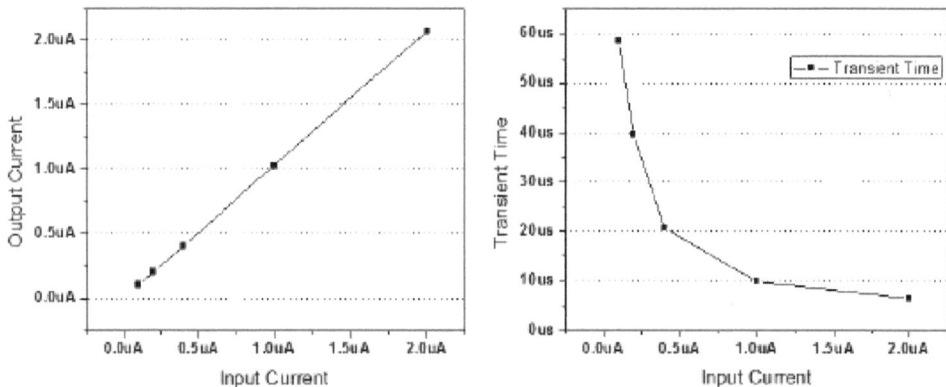

Fig. (6.25). Input-Output function and the data settling time of current driven 4T1C OLED pixel.

5T2C Pixel Configuration

Another OLED pixel employing voltage driven scheme is the 5T2C (five TFTs and two capacitors) configuration, proposed by Y.G. Mo *et al.* from Samsung Mobile Display [12, 13]. Fig. (**6.26**) shows the schematic of the threshold voltage shift compensated 5T2C OLED pixel. In compared to other designs, this pixel required pulsed common voltage (V_{COM}) and more control signals.

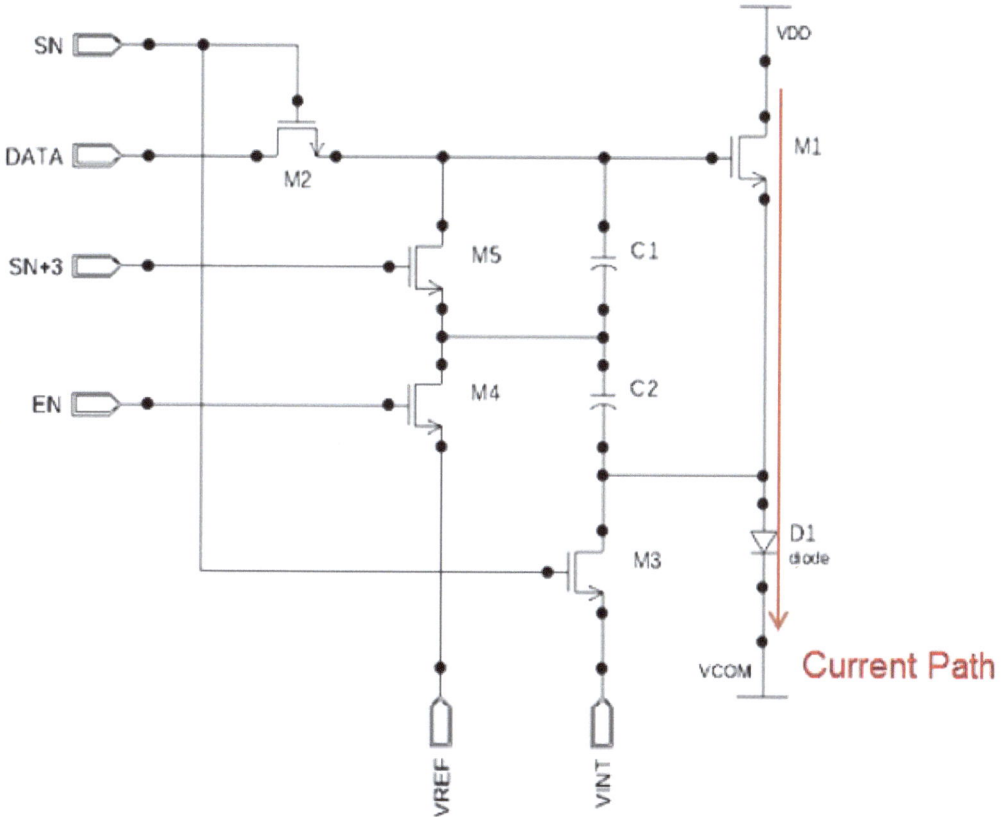

Fig. (6.26). Schematic of threshold voltage shift compensated 5T2C OLED pixel.

Fig. (**6.27**) sketched the driving waveform of threshold voltage shift compensated 5T2C OLED pixel. The operation of this pixel is divided into three stages. The first stage is the setup stage, in which all nodes in the pixel circuit are initialized. V_{COM} pulls high to prevent current from flowing through the OLED device.

Signal (EN) is on, allowing the DC reference voltage (V_{REF}) to connect to the capacitors C1 and C2. Signal (SN) asserts M2 and M3, so that the data signal and the initialization voltage (V_{INT}) can enter the pixel. Note that after initialization, V_{GS} of M1 is equal to $V_{DATA}-V_{INT}$.

The second stage is V_{TH} compensation stage; it is reserved for the threshold voltage extraction. In this stage, only the enable signal (EN) is on, keeping the existence of V_{REF} in between C1 and C2. At the same time, transistor M1 will charge up its source according to V_{DATA}, such that voltage at this node increase from V_{INT} to $V_{DATA}-V_{TH}(M1)$, where $V_{TH}(M1)$ is the threshold voltage of M1. This voltage is stored in C2 for the next stage.

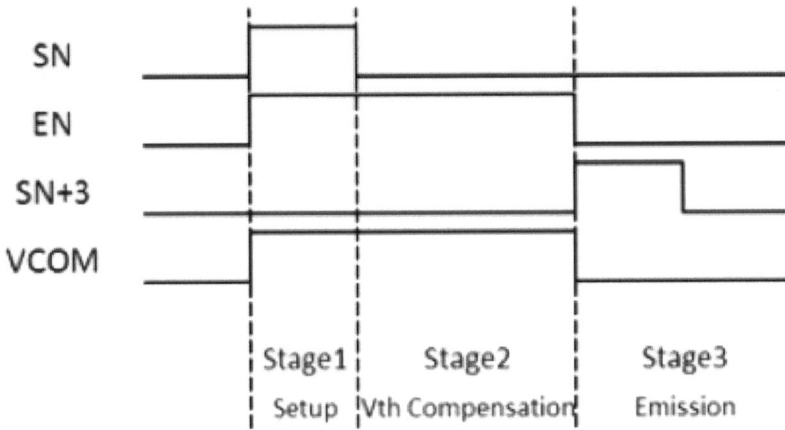

Fig. (6.27). Driving waveform of threshold voltage shift compensated 5T2C OLED pixel.

In the last stage, the pixel enters emission stage, in which the common voltage V_{COM} goes down, let current passes the OLED device for emission. Meanwhile, all signals pull low expect signal (SN+3), so transistor M5 is opened, the gate and the source of M1 become V_{REF} and $V_{DATA}-V_{TH}(M1)$ respectively. The V_{GS} of M1 and the pixel current, therefore, can be given by

$$V_{gs}= v_{REF} - V_{DATA} + V_{th(M1)}$$
$$I_{OLED}=K(V_{REF}-V_{DATA}+V_{th(M1)}-V_{th(M1)})^2=K(V_{REF}-V_{DATA})^2$$

where $K = \dfrac{\mu C_{ox}W}{2L}$. By the resultant pixel current, we can see that the current is no longer depended on the threshold voltage of M1. Thus, threshold-shift compensation is achieved. Fig. (**6.28**) shows the simulation result of pixels under threshold voltage shift, 5T2C exhibits better compensation ability, despite the complexity in the pixel circuit and driving signals.

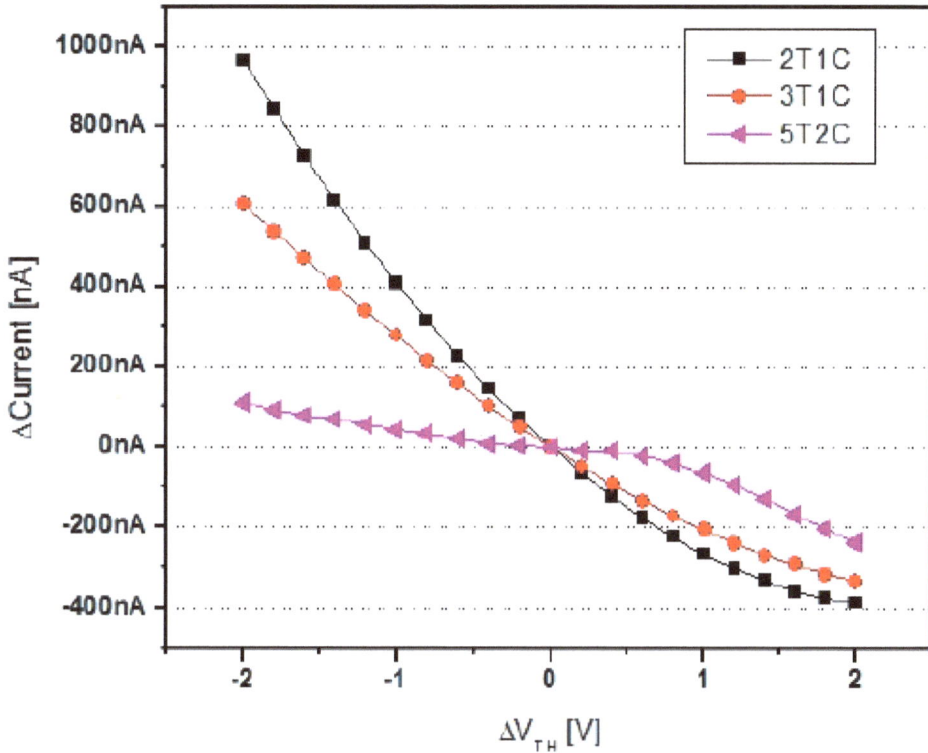

Fig. (6.28). Simulation result of current change against threshold voltage shift in 5T2C OLED pixel when the pixel current is 400 nA.

6T1C Pixel Configuration

The 6T1C pixel configuration is another threshold-compensated design proposed by Samsung Mobile Display [14]. Compared to the previous 5T2C design, the 6T1C version eliminates the need of switching common voltage (V_{COM}) and is driven with one less reference voltage. Fig. (**6.29**) shows schematic of the threshold-compensated 6T1C OLED pixel.

As seen from Fig. (**6.30**), the operation of the 6T1C pixel is divided to three stages, which are setup stage, threshold compensation stage and emission stage. First of all, this pixel is initialized through setup stage. Before it undergoes data update, the scan signal from previous row (SN-1) turns on transistor M4, which passes the initialization voltage V_{INT} into the pixel. This step effectively refreshes memory of the pixel by charging up capacitor C1 to V_{DD}-V_{INT}.

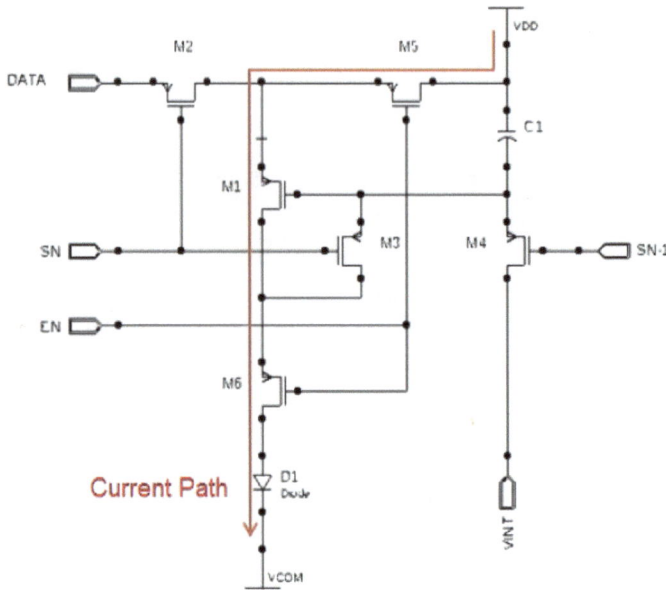

Fig. (6.29). Schematic of threshold voltage shift compensated 6T1C OLED pixel.

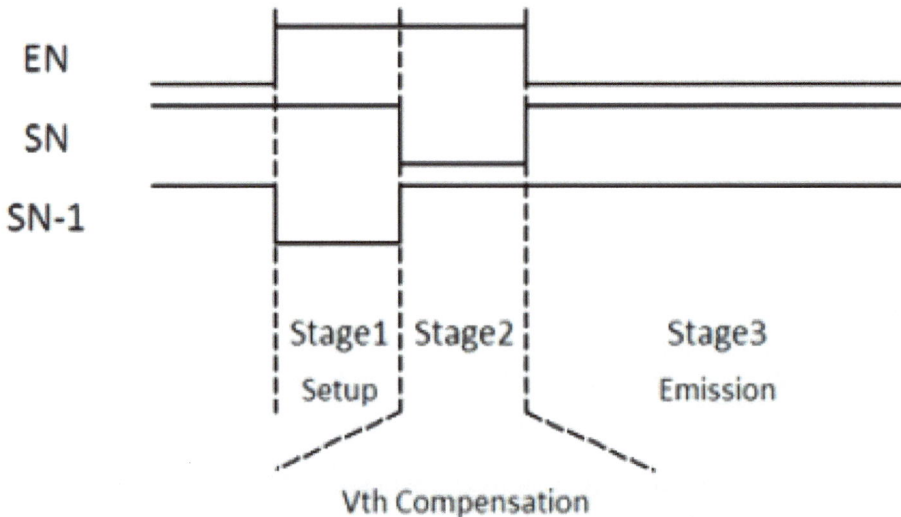

Fig. (6.30). Driving waveform of threshold voltage shift compensated 6T1C OLED pixel.

Then, the pixel enters compensation stage, where transistors M4, M5 and M6 are switched off. The scan signal (SN) leaves M2 and M3 transparent, M3 is then diode connects M1 and charge up C1 according to the data voltage (V_{DATA}), such that the potential at the source and the gate of M1 are V_{DATA} and V_{DATA}-V_{TH}(M1) respectively owing to the diode connection. This voltage, V_{DATA}-V_{TH}(M1), is stored to one end of the capacitor C1, with the other end connected to V_{DD}.

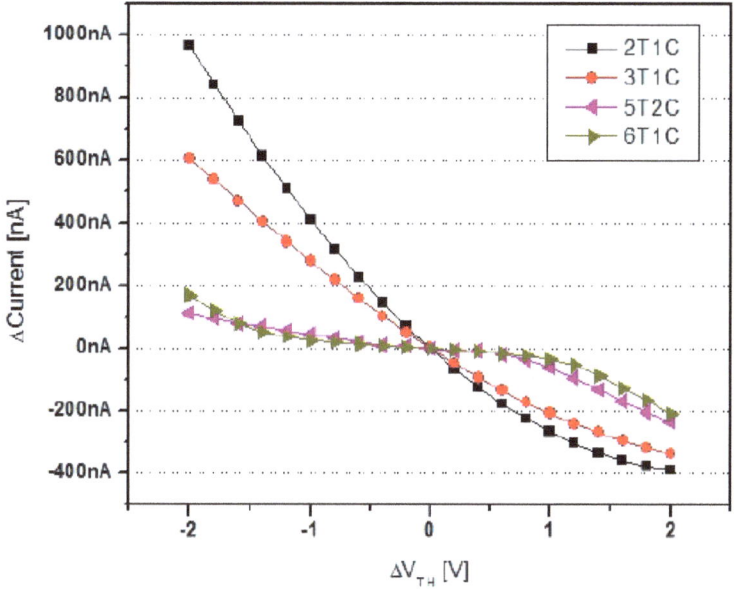

Fig. (6.31). Simulation result of current change against threshold voltage shift in 6T1C OLED pixel when the pixel current is 400 nA.

Finally, M2 and M4 are switched off in emission stage; only M5 and M6 are turned on by the enable signal (EN). In such connection, V_{GS} of M1 is just the stored voltage across C1, the pixel current can be given by

$$|V_{gs}|=V_{DD}-V_{DATA}+|V_{th(M1)}||$$
$$I_{OLED}=K(V_{DD}-V_{DATA}+|V_{th(M1)}|-|V_{th(M1)}|)^2=K(V_{DD}-V_{DATA})^2$$

The effective output current is, again, independent to the threshold voltage of the driving TFT, M1. The simulation result of current change against threshold voltage shift in 6T1C OLED pixel is shown in Fig. (**6.31**).

6T1C Pixel Configuration with biased discharge method

Depending on stressing condition, instability in threshold voltage can be observed in both positive and negative directions. Devices undergo more positively stresses may experience stronger positive-bias temperature instability (PBTI), which resulting in decease of absolute threshold voltage. Most threshold voltage shift compensated pixels use diode connection method to extract the device's threshold voltage. However, this can only handle half of the whole case. Fig. (**6.32**) indicates threshold voltage extraction using diode connected TFT under both positively and negatively shifted V_{TH}. When threshold voltage is positive, voltage at node A decrease from a pre-charged voltage to V_{TH}, in this case, V_{TH} is achieved. On the other hand, when the threshold voltage V_{TH} is negative, there is no way for the diode to extract such threshold because V_{GS} will not go below 0V, thus, the threshold information is lost.

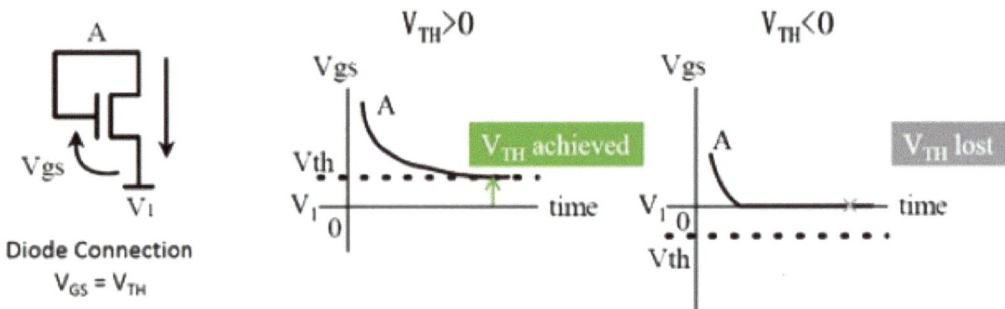

Fig. (6.32). Threshold voltage extraction through diode connection method.

Consider the extraction process with a slight change in the circuit, Fig. (**6.33**) depicts the circuit and signal of this new extraction method. In this method, given voltage V_3 is larger than $V_2 + V_{TH}$, the threshold voltage can be successfully extracted for both cases no matter the V_{TH} is positive or negative. Such threshold voltage extraction method is called the biased discharged method [15].

Using this extraction method, we have a new 6T1C threshold voltage shift compensated pixel. The pixel schematic and driving waveform is shown in Figs. (**6.34** and **6.35**). This pixel configuration is driven with four signals, one DC reference voltage and one power signal. The common voltage (VCOM) is fixed throughout the whole operation.

Fig. (6.33). Threshold voltage extraction through biased discharge method.

Fig. (6.34). Schematic of threshold voltage shift compensated 6T1C OLED pixel using biased discharge method.

Firstly, the operation of this pixel starts by initialization stage. The enable signals EN and V_{DD_EN} turn on transistor M2, M4, M5 and M6 that allow the initialization voltage (V_{INIT}) enters the pixel. This voltage prevent any current flowing through the OLED device and charge up capacitor C1 to V_{DD}-V_{INIT}.

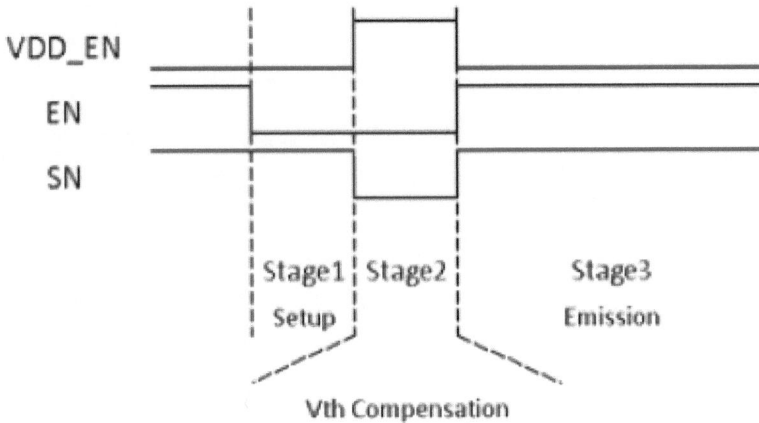

Fig. (6.35). Driving waveform of threshold voltage shift compensated 6T1C OLED pixel using biased discharge method.

Secondly, the pixel enters compensation stage, in which the biased discharge method is adopted. In this stage, all signals are on except V_{DD_EN}, therefore, the input data voltage biases the gate of M1; while the source of M1 goes down from V_{DD} to V_{DATA}+$|V_{TH}|$, and this voltage is stored in C1.

Finally, in emission stage, all signals are switched off except V_{DD_EN}, only transistors M2 and M5 are turned on, and the capacitor C1 is equivalently connected across the gate and source of M1. During the emission stage, the $|V_{GS}|$ of M1 and pixel current can be given by

$$|V_{gs}|=V_{DATA}+|V_{th(M1)}|$$
$$I_{OLED}=K(V_{DATA}+|V_{th(M1)}|-|V_{th(M1)}|)^2$$

From the simulation result of current change against threshold voltage shift of all compensated pixel as shown in Fig. (**6.36**). For 6T1C pixel adopting the biased discharge method, it shows better stability in pixel current and wider tolerable margin against threshold voltage shift.

In conclusion, threshold-compensated pixels generally show better compensation effect with more number of components used. This added feature, improved output stability against threshold voltage variation; on the other hand, increased pixel complexity and affected pixel aperture. Table **6.2** summarized the features of all discussed OLED pixel structures. In manufacturing consideration, having practical simplicity is as important as having good circuit features. The proposed 6T1C using biased discharge method takes balanced approach between these factors and can be a competitive candidate for future OLED display.

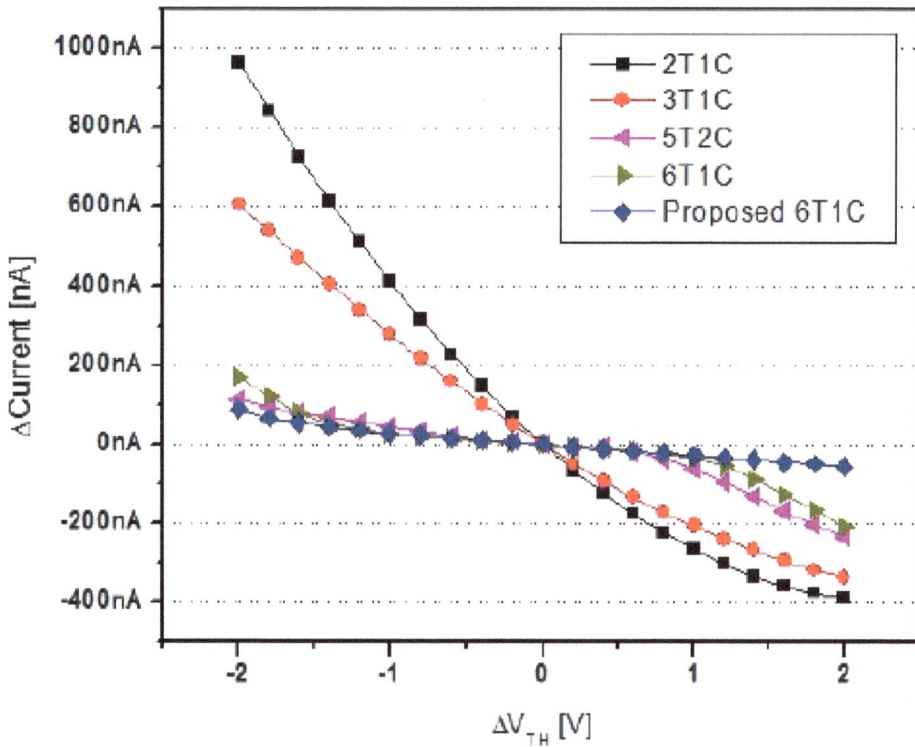

Fig. (6.36). Simulation result of current change against threshold voltage shift in 6T1C OLED pixel using biased discharge method when the pixel current is 400 nA.

Table 6.2. Table of comparison for features in different OLED pixel configurations.

	2T1C	3T1C	5T2C	6T1C	Propsed 6T1C
Compensation Ability	None	Fair	Good	Good	Excellent
Simplicity	Excellent	Good	Poor	Fair	Fair

(Table 6.2) contd.....

	2T1C	3T1C	5T2C	6T1C	Propsed 6T1C
Signal Compatibility	Excellent	Excellent	Poor	Fair	Fair
# Control Signals	2	2	4	4	4
# DC Ref and Power	1	1	3	2	2
Switching of DC power	No	No	Yes	No	No

CONFLICT OF INTEREST

The authors confirm that this chapter contents have no conflict of interest.

ACKNOWLEDGEMENTS

Declared none.

REFERENCES

[1]   http://hyperphysics.phy-astr.gsu.edu/hbase/electric/capchg.html..

[2]   J. M. Rabaey, A. P. Chandrakasan, and B. Nikolic, *Digital Integrated Circuits.* Prentice hall: Englewood Cliffs, 2002.

[3]   Z.D. Popovic, and H. Aziz, "Reliability and degradation of small molecule-based organic light-emitting devices (OLEDs)", *IEEE Journal Of Selected Topics in Quantum Electronics,* vol. 8, pp. 362-371, 2002.
        [http://dx.doi.org/10.1109/2944.999191]

[4]   D. Fish, M. Childs, S. Deane, J. Shannon, W. Steer, N. Young, A. Giraldo, H. Lifka, and W. Oepts, "Improved optical feedback for OLED differential ageing correction", *J. Soc. Inf. Disp.,* vol. 13, pp. 131-138, 2005.
        [http://dx.doi.org/10.1889/1.2012595]

[5]   D. Kondakov, W. Lenhart, and W. Nichols, "Operational degradation of organic light-emitting diodes: Mechanism and identification of chemical products", *J. Appl. Phys.,* vol. 101, p. 024512, 2007.
        [http://dx.doi.org/10.1063/1.2430922]

[6]   T. Tewksbury III, and H. Lee, "Characterization, modeling, and minimization of transient threshold voltage shifts in MOSFETs", *IEEE J. Solid-State Circuits,* vol. 29, pp. 239-252, 1994.
        [http://dx.doi.org/10.1109/4.278345]

[7]   S. Zafar, A. Callegari, E. Gusev, and M.V. Fischetti, "Charge trapping related threshold voltage instabilities in high permittivity gate dielectric stacks", *J. Appl. Phys.,* vol. 93, pp. 9298-9303, 2003.
        [http://dx.doi.org/10.1063/1.1570933]

[8]   K. Pernstich, S. Haas, D. Oberhoff, C. Goldmann, D. Gundlach, B. Batlogg, A. Rashid, and G. Schitter, "Threshold voltage shift in organic field effect transistors by dipole monolayers on the gate insulator", *J. Appl. Phys.,* vol. 96, pp. 6431-6438, 2004.
        [http://dx.doi.org/10.1063/1.1810205]

[9]   M. Exarchos, G. Papaioannou, D. Kouvatsos, and A. Voutsas, "An investigation of the electrically active defects in poly-si thin film transistors", In: *Thin Film Transistor Technologies (TFTT VII): Proceedings of the International Symposium.* The Electrochemical Society: New Jersey,USA, 2005, p. 125.

[10]   J. Lee, I. Cho, J. Lee, and H. Kwon, "Bias-stress-induced stretched-exponential time dependence of threshold voltage shift in InGaZnO thin film transistors", *Appl. Phys. Lett.,* vol. 93, p. 093504, 2008. [http://dx.doi.org/10.1063/1.2977865]

[11]   A. Shin, S.J. Hwang, S.W. Yu, and M.Y. Sung, "Design of organic TFT pixel electrode circuit for active-matrix OLED displays", *JCP,* vol. 3, pp. 1-5, 2008. [http://dx.doi.org/10.4304/jcp.3.3.1-5]

[12]   Y.G. Mo, M. Kim, C.K. Kang, J.H. Jeong, Y.S. Park, C.G. Choi, H.D. Kim, and S.S. Kim, "Amorphous-oxide TFT backplane for large-sized AMOLED TVs", *J. Soc. Inf. Disp.,* vol. 19, pp. 16-20, 2011. [http://dx.doi.org/10.1889/JSID19.1.16]

[13]   "Anonymous information display", *SID,* vol. 2, Mar-Apr 2013.

[14]   K. Kang, and N. Kim, *"Organic light emitting display having uniform brightness",* US 20120050344 A1, 2011.

[15]   T. Tanabe, S. Amano, H. Miyake, A. Suzuki, R. Komatsu, J. Koyama, S. Yamazaki, K. Okazaki, M. Katayama, and H. Matsukizono, "9.1: WITHDRAWN: 9.2: New threshold voltage compensation pixel circuits in 13.5-inch quad full high definition OLED display of crystalline In-Ga-Zn-Oxide FETs", *SID Symposium Digest of Technical Papers,* pp. 88-91, 2012. [http://dx.doi.org/10.1002/j.2168-0159.2012.tb05717.x]

SUBJECT INDEX